Intelligent Cities and Globalisation of Innovation Networks

Nicos Komninos

LONDON AND NEW YORK

First published 2008
by Routledge
2 Park Square, Milton Park, Abingdon, OX14 4RN

Simultaneously published in the USA and Canada
by Routledge
270 Madison Avenue, New York, NY10016

Routledge is an imprint of the Taylor & Francis Group, an informa business

© 2008 Nicos Komninos

Typeset in Bembo by
Florence Production Ltd, Stoodleigh, Devon
Printed and bound in Great Britain by
Cromwell Press, Trowbridge, Wiltshire

All rights reserved. No part of this book may be reprinted
or reproduced or utilised in any form or by any electronic,
mechanical, or other means, now known or hereafter invented,
including photocopying and recording, or in any information
storage or retrieval system, without permission in writing
from the publishers.

British Library Cataloguing in Publication Data
A catalogue record for this book is available from the British Library

Library of Congress Cataloging in Publication Data
Komninos, Nicos.
  Intelligent cities and globalisation of innovation networks/
  Nicos Komninos.
     p. cm. – (Regions and cities)
   Includes bibliographical references and index.
   1. City planning – Technological innovations. 2. Regional
planning – Technological innovations. 3. Information
networks – Social aspects. 4. Communication – Technological
innovations. 5. Globalisation – Economic aspects. I. Title.
   II. Title: Intelligent cities and globalisation of innovation networks.
HT166.K626 2008
307.1′216–dc22   2007048599

ISBN10: 0–415–45591–X (hbk)
ISBN10: 0–415–45592–8 (pbk)
ISBN10: 0–203–89449–9 (ebk)

ISBN13: 978–0–415–45591–6 (hbk)
ISBN13: 978–0–415–45592–3 (pbk)
ISBN13: 978–0–203–89449–9 (ebk)

Intelligent Cities and Globalisation of Innovation Networks

The development of a city in the twenty-first century is at the crossroad of the rising knowledge economy and the increasing digitalisation of contemporary urban life. *Intelligent Cities and Globalisation of Innovation Networks* outlines a new paradigm of urban and regional development, emerging from the meeting of the knowledge economy with the virtual world and the global outsourcing networks.

This book assesses the evolution and continuous spatial enlargement of territorial systems of innovation under the pressure of globalisation, and the rise of intelligent environments (clusters, cities, regions) as a means of dealing with global information networks, global new product development, and global knowledge outsourcing. It also looks at the knowledge functions of intelligent cities (strategic intelligence, technology absorption, cooperative new product development, global supply chains and digital-city marketplaces) as essential building blocks of these cities.

It asks key questions of the movement towards intelligent environments and cities, such as:

- What makes an intelligent city and how do we define it?
- What makes some cities and regions more efficient in developing knowledge, technologies and innovations?
- How important is the external urban environment to the innovation performance of an organisation?
- How is innovation changing along with digitalisation and globalisation?

This book will be essential reading for urban and regional planners, development consultants, and innovation professionals, and will be of interest to students and researchers in the field.

Nicos Komninos is currently a professor of Urban Development and Innovation Policy at the Aristotle University of Thessaloniki. He is Director of the URENIO Research Unit and has co-ordinated numerous research projects under the European R&D Framework Programmes. He has been involved in the development of technology parks and regional innovation strategies in the EU and developing countries.

Regions and cities
Series editors: Ron Martin, University of Cambridge, UK,
Gernot Grabher, University of Bonn, Germany,
Maryann Feldman, University of Georgia, USA

Regions and Cities is an international, interdisciplinary series that provides authoritative analyses of the new significance of regions and cities for economic, social and cultural development, and public policy experimentation. The series seeks to combine theoretical and empirical insights with constructive policy debate and critically engages with formative processes and policies in regional and urban studies.

Intelligent Cities and Globalisation of Innovation Networks
Nicos Komninos

Devolution, Regionalism and Regional Development
Jonathan Bradbury (ed)

Creative Regions
Technology, culture and knowledge entrepreneurship
Philip Cooke and Dafna Schwartz (eds)

European Cohesion Policy
Willem Molle

Geographies of the New Economy
Peter Daniels, Michael Bradshaw,
Jon Beaverstock and Andrew Leyshon (eds)

The Rise of the English Regions?
Irene Hardill, Paul Benneworth, Mark Baker and Leslie Budd (eds)

Regional Development in the Knowledge Economy
Philip Cooke and Andrea Piccaluga (eds)

Clusters and Regional Development
Critical reflections and explorations
Bjørn Asheim, Philip Cooke and Ron Martin (eds)

Regions, Spatial Strategies and Sustainable Development
Graham Haughton and Dave Counsell (eds)

Geographies of Labour Market Inequality
Ron Martin and Philip Morrison (eds)

Regional Development Agencies in Europe
Henrik Halkier, Charlotte Damborg and Mike Danson (eds)

Social Exclusion in European Cities
Processes, experiences and responses
Ali Madanipour, Goran Cars and Judith Allen (eds)

Regional Innovation Strategies
The challenge for less-favoured regions
Kevin Morgan and Claire Nauwelaers (eds)

Foreign Direct Investment and the Global Economy
Nicholas A. Phelps and Jeremy Alden (eds)

Restructuring Industry and Territory
The experience of Europe's regions
Anna Giunta, Arnoud Lagendijk and Andy Pike (eds)

Community Economic Development
Graham Haughton (ed)

Out of the Ashes?
The social impact of industrial contraction and regeneration on Britain's mining communities
David Waddington, Chas Critcher, Bella Dicks and David Parry

Contents

List of illustrations ix
Acknowledgements xi

Introduction 1

PART 1
Globalisation of innovation and intelligent cities 5

1 **An intelligent global world is emerging** 7

 A new global setting 7
 Globalisation of innovation 8
 Europe in search of innovation 11
 Innovation reshaping core and periphery 13
 Innovation turning a new leaf for cities 15
 Trading innovation with intelligent environments 16
 Intelligent environments 17

2 **Regions of innovation excellence: learning from the best** 21

 A new model of regional development 21
 The rise of the innovation economy in the EU 23
 Regions of innovation excellence in Europe 30
 The core of excellence: regional systems of innovation 33
 Making the core of excellence: four stages of evolution 38

3 **Systems of innovation: continuous spatial enlargement** 45

 Widening systems of innovation 45
 Innovation 47
 Towards systems of innovation 49

What makes an innovation system? 54
Enlargement I: cluster-type innovation systems 56
Enlargement II: regional systems of innovation 64
Enlargement III: intelligent global systems of innovation 71
Governance of territorial systems of innovation 73

4 Virtual innovation environments: enriching innovation systems with global networks and users participation — 77

Innovation as an environmental condition 77
Virtual innovation environments 80
Functioning of VIEs 89
Reception of VIEs: lessons from the 'VERITE' network 91
'VERITE' as a virtual community 95
Learning impact 100
Direct impact 102
Innovation on virtual space 106

5 Intelligent cities: the emergence of the concept — 110

Intelligent cities: bridging physical and virtual worlds 110
Cyberspaces and cyber cities 113
Smart communities 116
Intelligent communities 117
Intelligent environments of innovation – Living Labs 119
Defining intelligent cities 120
Three layers of intelligent cities 123
A glimpse at intelligent cities around the world 125
Metrics 131

PART 2
Building blocks of intelligent cities — 135

6 Strategic economic intelligence: global watch on markets and technologies — 137

Innovative regions in search of intelligence 137
Strategic economic intelligence: business, cluster, and regional 140
Main components of strategic economic intelligence 149
Integration of distributed intelligence 156
Meta-foresight: an experiment in strategic economic intelligence 160
The Meta-foresight platform 161
Sectoral/cluster intelligence 166
Strategic business intelligence 168
Strategic economic intelligence: the search for the 'next big thing' 172

7 Technology transfer and acquisition: virtual spaces valorising academic research 174

Getting state-of-the-art technology 174
The landscape of technology transfer 175
The turn towards academia 177
Technology transfer as communication 180
Blending cooperation networks and virtual spaces 181
The DRC: a virtual space for the valorisation of academic research 183
Augmented technology transfer 192

8 Innovation through collaboration: managing networks that cross boundaries 195

Innovation through collaboration 195
The logic of collaborative innovation networks 196
Selection of partners: trust vs. competence 200
Spatiality of collaboration: contact vs. cost 202
Digital spatiality and networking 204
Managing collaborative innovation networks: three experiments 205
InnoCentive: innovation based on global networks 205
Regional Innovation Poles: multilevel regional innovation networks 208
New product development based on vertical networks 216
Collaborative innovation as intelligent space 222

9 Digital cities and e-marketplaces: global promotion of localities, products and services 226

The last mile of innovation 226
Digital cities: broadband networks and online services 227
Broadband network 230
Online services 233
e-City marketplaces 235
e-City promotion 240
e-Government 242
Sustainability of digital cities and e-marketplaces 245

10 Building blocks of intelligent cities: architecture of layers and functions 247

Two movements shaping intelligent cities 247
Intelligent cities as territorial systems of innovation 249
Intelligent cities as digital spaces of collaboration 257

Integration: knowledge functions of intelligent cities 261
The architecture of intelligent cities 267
Intelligent cities: a window to global innovation networks 270

Appendix: five platforms for intelligent cities	275
Notes	283
References	284
Index	301

Illustrations

Figures

2.1	Innovation performance in EU Member States	26
2.2	Innovation performance in EU Member States 2002–2006	27
2.3	Global Innovation Index	28
2.4	Regional disparities in innovation performance	28
2.5	In-house innovation	34
2.6	Innovation within the regional system of innovation	36
3.1	Knowledge evolution along the innovation process	55
3.2	Cooperation circuits into regional systems of innovation	65
3.3	Innovation policy adapted to regional systems of innovation	66
3.4	Precarious regional systems of innovation	68
4.1	Structure of web-based intelligence	83
4.2	Online innovation management tools: overview	87
4.3	Homepage of VERITE	96
4.4	VERITE services & tools	98
4.5	VERITE knowledge base	99
4.6	Logical impact of VERITE	100
4.7	Motivation for joining VERITE	101
4.8	Frequency of use of the virtual tools	104
4.9	Relevance of digital tools of VERITE	105
6.1	Integration of informational modules	158
6.2	Meta-foresight structure	162
6.3	Meta-foresight homepage	163
6.4	Meta-foresight toolbox	164
6.5	Meta-foresight portal	167
6.6	Data entry and analysis tool	170
6.7	Data analysis: Eight themes under seven perspectives	171
7.1	The landscape of technology transfer	176
7.2	Technology transfer as communication	181
7.3	DRC architecture – physical and virtual components	184
7.4	DRC layer 1 – a database of technologies and expertise	186
7.5	DRC layer 2 – structure of spin-off creation roadmap	187
7.6	DRC layer 3 – technology dissemination	188

8.1	Innovation network actors and roles	198
8.2	Nodes within an innovation network	203
8.3	Innovation system in the ICT sector	212
8.4	NPD-Net concept	217
8.5	Online roadmap of new product development	219
9.1	Digital city network and services	228
9.2	Layers of digital city services	235
9.3	AOL Digital cities	237
9.4	Corfu – digital city and e-market place	238
9.5	Virtual Canberra	242
10.1	Knowledge networks in product innovation	252
10.2	Knowledge networks in technology transfer processes	255
10.3	Knowledge networks within the supply chain	256
10.4	Digital cities levels	260
10.5	A digital city as sum of all webs for the city	262
10.6	Architecture of intelligent cities	269
A.1	Five platforms for intelligent cities	276
A.2	Strategic economic intelligence platform	278
A.3	Technology dissemination platform	279
A.4	Collaborative innovation platform	281
A.5	New company incubation platform	282
A.6	Virtual tour and e-marketplace platform	283

Tables

1.1	Business R&D concentration and intensity	9
2.1	Regions of innovation excellence in the EU	31
2.2	Dimensions and typology of regional clusters	40
3.1	Evaluation of innovative start-ups	63
3.2	Advantages and challenges in territorial systems of innovation	75
4.1	Tools for making virtual innovation environments	81
4.2	Four building blocks of VERITE	92
5.1	Top intelligent communities selected by the ICF 2001–2007	126
5.2	Metrics for intelligent cities	133
6.1	Business, cluster and regional intelligence	144
6.2	Data collection: questions per subject and source	169
7.1	Changing technology transfer routes (EU-15)	178
8.1	Type of networks and degree of member integration	200
8.2	Collaboration mechanisms within Regional Innovation Poles	210
9.1	Broadband networks and technologies	231
9.2	Characteristics of xDSL networks	232
10.1	Intelligent cities building blocks and knowledge functions	272

Acknowledgements

Since the publication of *Intelligent Cities: Innovation, knowledge systems and digital spaces*, in 2002, I have been involved in a series of research and planning projects that are behind the concepts and ideas contained in this book. I have enjoyed the cooperation and valuable contribution from numerous individuals who took part in this multifaceted research on innovation and intelligent environments in the context of projects such as VERITE, MEDICUBE, K-CLUSTERS, INVENT, METAFORESIGHT, STRATINC, NETFORCE, and FUTURREG. I am indebted to all of them, although it is not possible to mention each one by name in this note.

Intelligent Cities and the Globalisation of Innovation Networks continues in the same direction taken by the book from 2002. However, this book codifies a research agenda on the environment of innovation that takes a step further from cluster strategies and regional innovation strategies. In this sense, *Intelligent Cities* is a 'third generation' strategy for making environments of innovation (after innovative clusters and regional systems of innovation), taking into account the current globalisation of innovation networks and the advantages offered by information and communication technologies. It is a strategy that focuses on the knowledge functions of cities and regions, which enable them to obtain strategic information, state-of-the-art technology, innovation through cooperation, and globally promote products and services.

Among the people I would like to thank are my colleagues at the URENIO Research Unit, especially Panagiotis Tsarchopoulos and Isidoros Passas who helped me with the development of prototypes for digital cities and virtual innovation spaces; Pierre Bourgogne for sharing his insights about the amazing world of strategic economic intelligence; and the late Philippe de Lavergne, a gifted person and an expert in regional development and innovation management, with whom I worked on the Meta-foresight project. I have profited enormously from the discussions and applications presented at two conferences on *Intelligent Environments*, IE 06 in Athens and IE 07 in Ulm. I am indebted to John A. O'Shea for undertaking the difficult task of editing the second draft of the book and navigating the waters of urban development, innovation, and virtual environments. I would like also to thank Caroline Mallinder, Georgina Johnson and the editorial staff at Routledge for their support in the production of this book. Finally I would like to thank my family, Elena and Alex, for the support and encouragement I received during the years required to write this book.

Permissions are gratefully acknowledged from the following: Alasdair Reid for Figure 4.6; AOL for Figure 9.3. AOL and the AOL Triangle Logo are registered trademarks of AOL LLC. © 2007 AOL LLC. MapQuest and the MapQuest logo are registered trademarks of MapQuest, Inc. Map content © 2007 by MapQuest, Inc and NavTeq. Used with permission. The Australian Government for Figure 9.4; Figure 10.5 is produced by the Mapstan virtual machine (www.mapstan.com).

October 15, 2007

Introduction

This book is about innovation, cities and regions. However, it approaches the subject via a substitution. The central premises on which our arguments are based are that (a) a theory of innovation is not feasible, since its prevision capacity would annihilate innovation itself, and (b) the quest for a theory of innovation can be replaced by the quest for understanding the environment for generating innovation. The substitution consequently lies in the assumption that we cannot predict the emergence of innovation and consequently manage it, but we can create environments within which innovation is generated; in other words that we can manage the environment rather than innovation itself. The more radical and disruptive innovation is, the more this substitution is necessary.

This is the central idea behind intelligent cities. The objective is about innovation, but the means which permit us to reach it are intelligent cities and other forms of intelligent environments sustaining the processes of innovation, which are now global.

So, the book starts as a book about innovation and the turn towards global innovation networks, but soon becomes a book about intelligent cities, communities, and clusters. It explains why intelligent environments are important today; sets out their role within global innovation networks; and discusses how we can create such environments.

Some of the fundamental questions the book tries to answer are:

- How important is the external environment to the innovation behaviour and capability of an organisation? Can an organisation go beyond the knowledge assets of its local environment? How is innovation changing along with globalisation?
- What makes some territories more efficient in developing technologies and innovations? What makes some territories more efficient in increasing the capabilities of the organisations located within them?
- What is the meaning of 'intelligent cities'? What makes an intelligent city? Is it just a question of technology? Can we give concrete examples of intelligent cities? Is the term a metaphor or we can speak literally about intelligent environments and cities?

2 *Introduction*

The book is not a narrative one. It was not written to narrate the story of the birth of intelligent cities, but to tell how new ideas and innovative products emerge within intelligent environments, clusters, and cities. From the outset, the writing of the book encountered an internal problem of understanding intelligent cities; how to figure out their meaning, literally or metaphorically; how they are composed; how they operate; how we can measure their performance. Thus writing the book was a process of research, which records assumptions and solutions and uses that record as a means for understanding and improving the ideas it was developing.

When work on the book began back in September 2002 the main intellectual constructs which are described in its ten chapters did not exist. The initial formulation of the core question in the book concerned 'innovation as an environmental condition', or the relationship of an innovative organisation to its physical and virtual environment. How determinative is that relationship for new knowledge creation and the knowledge capacity of the organisation? How important is the impact of the external environment on the innovation performance of the organisation? Can an organisation exceed its knowledge boundaries and skills imposed by its immediate environment? Can internal procedures for generating knowledge overturn the research orientations coming from the external environment? What is the relationship between the generating mechanisms of strategic intelligence, technology transfer, and cooperative innovation?

In the process of writing, the core concepts that outline 'the intelligent city paradigm' and run through this book were formed. Ideas such as:

- the restructuring and widening of territorial systems of innovation with respect to the globalisation of innovation networks and the intense use of information and communication technologies;
- intelligent cities as advanced territorial systems of innovation integrating innovation processes and digital collaboration spaces;
- innovative clusters as the core elements of intelligent cities and the latter as complex collectives of clusters;
- intelligent cities as a synthesis of the physical, institutional, and the digital spatiality of the innovation process;
- intelligent cities as a synthesis of intellectual capital, social capital, and information technology applications in the field of innovation;
- strategic intelligence, technology acquisition, cooperative innovation, and global promotion as key knowledge functions within intelligent cities; and
- planning of intelligent cities based on digital platforms sustaining the knowledge functions of these cities.

Writing encompassed creation, selection, and synthesis of ideas. There was no initial model of an intelligent city to follow (and thus no story to narrate); rather a series of questions about its formation and operation that received the answers provided here. I consider that by the end of this book readers will have a well-rounded picture about what intelligent cities are, how they operate, how they can be designed today, and the limits on knowledge and technology we have in making intelligent cities.

The greater part of the book is based on applied research. The arguments presented in its ten chapters and the digital applications described, document and correlate applied research projects funded by EU research programmes over the period 2000–2007. More specifically:

- The analysis contained in Chapter 2 is based on research into measurement and benchmarking of innovation systems in the European regions (INNOREGIO, 2000–2002 and EMERIPA, FP6 2005–2006). An early form of this chapter appeared in Komninos N. (2004) 'Regions of Excellence in the EU: A new model of regional hierarchy and development', in G. Kafkalas (ed.) *Spatial Development Issues*, Athens, Kritiki Press (in Greek).
- Chapter 4 is based on the experiences and the conclusions of the VERITE project (FP5, 2002–2003).
- The organisational layout and digital platform described in Chapter 6 is based on the conclusions of the Meta-foresight (Regions of Knowledge, 2004–2005) project. Part of it was published as: Komninos, N. (2004) 'Regional intelligence: Distributed localized information systems for innovation and development', *International Journal of Technology Management*, Vol. 28, No 3-4-5-6, pp. 483–506.
- The case study and digital solution referred to in Chapter 7 is based on the application developed by the Virtual Research Centre (Innovative Actions Programme, 2003–2005). An early form of this paper was published at the conference proceeding 'IE 06': Komninos, N., Sefertzi, E., and Tsarchopoulos P. (2006) 'Virtual innovation environment for the exploitation of R&D', *Intelligent Environments 06*, Institution of Engineering and Technology, pp. 125–36.
- Chapter 8 is based on the conclusions and the knowledge developed in the context of the NPD-NET, an Interreg 3C project (2004–2006).
- Chapter 9 refers to and outlines insights gained by the development of digital cities (Digital Corfu and Digital Aegean) in the framework of the respective Innovative Actions Programmes (2003–2005).

The book's structure is simple. Its two parts examine the emergence and functioning of intelligent cities from two different perspectives.

The first part of the book is about the rise of intelligent cities as a major paradigm of urban development and planning in the twenty-first century. It looks at the evolution of territorial systems of innovation under the pressure of globalisation and the rise of intelligent environments (cities, clusters, regions) as a means of dealing with global innovation networks, global new product development, and global knowledge outsourcing. It starts from the regions of innovation excellence in Europe and the lessons we can learn from them; goes through the description of different forms of territorial systems of innovation; and explains the added value of digital networking and online interaction within innovation systems. This part ends with an overview of the movements for making intelligent cities; a quick look at cities in the US, Europe and Asia that have been recognized as intelligent, and advances a definition of the fundamental elements of an intelligent city.

The second part is about the knowledge and innovation functions within intelligent cities. It starts with a description of the main functions of intelligent cities, such as strategic intelligence, technology acquisition, cooperative product development, global supply chains and digital-city marketplaces. Each function is sustained by networks of creative people and knowledge-intensive organisations, institutions for managing knowledge and innovation, and digital networks and online services. The mixture is the motor of the city's intelligence as it brings together human skills, technology learning institutions, and digital spaces for learning and cooperation. This part ends with a description of the building blocks of intelligent cities: the three levels that make an intelligent city (creative people and organisations, innovation support institutions, digital innovation spaces) and the four functions that enable people to work cooperatively and master different forms of knowledge and know-how.

The book combines concepts and theories from three different fields of science and technology: (1) urban development and planning; (2) innovation management; and (3) virtual/intelligent environments. The main features of this work are that it explains the rise of intelligent cities with respect to the globalisation of systems of innovation; it opens up a new way for making intelligent environments via the connection of human skills, institutional mechanisms, and digital spaces operating within a community; and it offers a series of digital platforms and tools for the making of intelligent cities.

Within the wider literature about cyber- and intelligent cities the book sides in favour of intelligent communities. The difference is that while cyber-cities mainly focus on technologies (communication networks, sensors, intelligent agents, automation of collection and information management) and the digital infrastructure of cities, intelligent communities emphasise the human, institutional, and digital aspects of agglomeration, as they emerge from the integration of human creativities, cooperation in innovation, and artificial intelligence applications available within a community. Intelligent cities correspond to a new type of agglomeration emerging out of innovative milieux, reflexive institutions of innovation, and interactive online services operating within global supply chains.

Part 1
Globalisation of innovation and intelligent cities

1 An intelligent global world is emerging

A new global setting

The world is changing. All we know about technology, production, trade, creation and the distribution of wealth is becoming rapidly obsolete. A new world is rising: a global world fuelled by information technologies, knowledge flows, innovation networks, and global supply chains. A new generation of cities and regions is rising also: knowledge-intensive, innovative and intelligent.

The beginning of the twenty-first century is marked by a major turn of the West towards the knowledge-based economy. In leading countries and regions in the Western world (US, Japan, European Union), competition and growth are taking place mainly in terms of R&D and technological innovation, while most dynamic sectors of industry draw their competitive advantage from knowledge, research and innovation. Europe has fully embraced the target of becoming the most competitive and dynamic knowledge-based economy in the world by 2010. Although it quickly became clear that this target was not realistic, it does mark the orientation of the EU towards a new model of development and prosperity based on knowledge, technology, and innovation.

The path to a knowledge-based economy has been accelerated by the globalisation of the economy and capital accumulation. Global flows of goods and knowledge-intensive services, global supply chains, and global research networks are becoming the new milestones in developed countries. In the core regions remain mainly activities of the new economy, knowledge-intensive manufacturing (computers, telecommunications, pharmaceuticals, aeronautics, and other), and knowledge-intensive services (IT, health, business, and financial services). Traditional industries, such as textiles, clothing and footwear, metals, shipbuilding, electrical appliances, with highly standardised work practices and low know-how are gradually moving to developing regions. This new division of labour with further segregation of manufacturing and services and the location of segments of production all over the world has resulted in the carriage of goods skyrocketing, creating major needs for global supply chain management. In almost all sectors of the economy, production has become global with research, design and development of new products taking place in the developed world and in a small number of selected metropolises and innovative hubs in developing countries, industrial mass production in the developing world, and consumption again at the industrial core. The need to coordinate the production cycle, research, services, manufacturing and consumption on a global scale is now a top priority, as is the need for relentless communication and exchange of information in real time round the clock. In global production landscapes, the sun never sets. Work in the fields of services and processing complement each other in the northern and southern hemispheres, day and night, without break, under continuous coordination and control.

Knowledge and innovation have become the golden keys for managing and controlling the global economy today. Key functions of technological intelligence, technology transfer, and innovation are built on global networks and on communications technologies that bring capabilities and creativities from around the world into partnership with each other.

However, this new setup is highly conflicting and unstable. In the developing world, which offers abundant cheap labour, a local technological base is gradually developing with increasing levels of complexity and ability. Since the knowledge-based economy is a human-centred economy, population-intense countries like Brazil, Russia, India, and China have a strong comparative advantage: abundant human resources and talent deriving from the statistics of their large populations. Resource, skill and technology concentration in the developing world is changing the established global equilibrium among Asia, Europe and the US, and is pushing the core regions to further strengthen their technological base and innovation capability.

The new global setting goes hand-in-hand with a new development triad: knowledge-based and innovation-led economies, supra-national regulation, and intelligent agglomerations. This new holy triad brings in new conflicts and unevenness. But, at the same time a more open society is emerging, since human skills and abilities from around the whole world can be valorised without prejudice based on skin colour, race or nationality.

Globalisation of innovation

For two consecutive years Booz Allen Hamilton (BAH), a leading global strategy and technology consulting firm, published the 'Global Innovation 1000' study, based on the 1,000 companies that were identified as the world's largest R&D spenders in 2004 and 2005. The study made an effort to assess the influence of R&D and innovation on corporate performance, analysing statistical relationships between R&D spending and primary indicators of company success, such as sales growth, gross margin percentage, gross profit growth, operating margin percentage, operating income, total shareholder return, and market capitalisation growth (Jaruzelski et al. 2006). The primary conclusion was that there is no simple relationship between R&D spending and corporate profitability:

> Money simply cannot buy effective innovation. There are no significant statistical relationships between R&D spending and the primary measures

of financial or corporate success: sales and earnings growth, gross and operating profitability, market capitalisation growth, and total shareholder returns. Gross profits as a percentage of sales is the single performance variable with a statistical relationship to R&D spending.

(Jaruzelski et al. 2006)

This should be expected however; funding alone cannot secure innovation: it needs the mobilisation of a wider set of human and institutional resources. However, what really stands out in this study is both the sectoral concentration and globalisation of R&D activity.

Among the top 1,000 R&D spenders, three industry sectors spent nearly two-thirds of the total R&D budget: computing and electronics (26 per cent), health (22 per cent), and automotive (17 per cent). Two of them (computing and electronics, and health) were also the most intense R&D performers in 2005, as shown by their high ratio of R&D-to-sales. Software and Internet industries, though spending less than 5 per cent of the total R&D expenditure, were also intense R&D performers, holding the first and second position in 2004 and 2005 respectively in the R&D-to-sales ratio.

However, the intensity of innovation, as measured by the R&D-to-sales ratio, in the period 2001–2005 has declined. Sales are rising, R&D spending is rising also, but the R&D-to-sales ratio is steadily declining from 4.09 per cent in 2001 to 3.84 per cent in 2005. The study argues that much of this decline can be explained by the increasing globalisation of R&D, which is outsourced to facilities in lower-cost regions of the world. The majority of the new R&D centres that companies plan to open during the next few years are to be located in India and China. The heavy concentration of R&D in the three industries makes its geographical distribution extremely dependent on the location and the outsourcing choices of these industries too.

Table 1.1 Business R&D concentration and intensity

Industries	R&D spending (% of total)	Rank	R&D-to-sales ratio (%)	Ranking
Computer and Electronics	26.0	1	7.5	3
Health	22.0	2	11.5	1
Auto	17.0	3	3.9	6
Chemicals and Energy	7.0	4	1.1	10
Technology	6.5	5	4.0	5
Industrials	5.7	6	2.2	7
Software and Internet	4.9	7	11.2	2
Aerospace and Defence	4.0	8	4.0	4
Consumer	3.8	9	2.0	8
Telecom	1.3	10	1.5	9
Other	2.5		3.84	

Source: Based on Jaruzelski et al. (2006).

The geography of business R&D expenditure is also radically changing. The BAH study looking at data for 2000–2005 shows that spending in developing countries is growing at incredible rates. While business R&D expenditure rose by 5.2 per cent in North America, 2.3 per cent in Europe, and 3.8 per cent in Japan during 2000–2005, it rose by 17 per cent in India and China, and by 19.7 per cent in Australia, Brazil, Singapore, South Korea, and Taiwan combined. These growth figures represent performances that are beyond any reach and imagination in the so-called first world.

Another source of statistical documentation, the Science and Technology Indicators of CORDIS (STI-ERA) confirm the same trends (Cordis 2007). The figures for 2005 show that EU R&D intensity measured by the R&D-to-GDP ratio has been slowing down since 2000, and grew only 0.2 per cent in the period 2002–2003. The opposite is taking place in China, which has lower R&D intensity (1.31 per cent of GDP in 2003), but with a fast growing rate of about 10 per cent per year between 1997 and 2002. If these trends continue, China will be spending the same as the EU in 2010. One of the reasons for this has been the redirection of business R&D funding. US investment has been growing at a much greater rate in areas outside the EU, about 8 per cent per year in the EU and 25 per cent per year in China. At the beginning of the 1990s, nearly 80 per cent of the total US overseas R&D investment was made in the EU, while at the end of the same decade this investment was reduced to 70–2 per cent. The decline in Europe was counterbalanced with the sharp increase in other regions, mainly China, where US R&D expenditure increased twenty-fold from 5 ME to 120 ME in ten years.

What do these trends signify? There is no doubt that a series of regions in the developing world have become extremely attractive for EU and US business R&D. Cisco already has R&D facilities in Bangalore, as does Toyota in Thailand. Nokia operates nine satellite design studios located within targeted nations like India (Bangalore), China (Beijing), and Brazil, where researchers and designers work to customise products to each market (BusinessWeek 2007). Companies are attracted by the low cost of labour, the culture of work, loyalty of the workforce, and the gradual opening up of huge local markets. In fact, what we observe behind the figures is the globalisation of innovation networks and the extension of product development cycle on a global scale. Thomas Friedman summarised these trends very illustratively in his best seller on our 'Flat World'. Every product, he writes:

> from software to widgets – goes through a cycle that begins with basic research, then applied research, then incubation, then development, then testing, then manufacturing, then deployment, then support, then continuation engineering in order to add improvements. Each of these phases is specialised and unique, and neither India nor China nor Russia has a critical mass of talent that can handle the whole product cycle for a big American multinational. But these countries are steadily developing their research and development capabilities to handle more and more phases.

As that continues, we really will see the beginning of what Satyam Cherukuri, of Sarnoff, an American research and development firm, has called 'the globalisation of innovation' and an end to the old model of a single American or European multinational handling all the elements of the product development cycle from its own resources. More and more American and European companies are outsourcing significant research and development tasks to India, Russia, and China.

Friedman (2006; pp. 29–30)

Globalisation of innovation comes together with the rediscovery of the East. Companies trade innovation analytical methods against extraordinary human skills based on the statistics of large populations. The knowledge-based economy is a human-centred economy after all; and the large population of giant countries of the East offers an indisputable advantage. BusinessWeek Magazine (2005) rightly pointed out that what was once central to corporations – price, quality, and much of the left-brain, digitised analytical work associated with knowledge – is fast being shipped off to lower-paid, highly trained Chinese and Indians, as well as Hungarians, Czechs, and Russians: 'The game is changing. It isn't just about math and science anymore. It's about creativity, imagination, and, above all, innovation.'

Europe in search of innovation

In innovation and knowledge-based development, the EU is still lagging behind. Due mainly to recent geographical enlargements, the gap between the EU, the US, and Japan is widening.

Using a set of twelve comparable indicators, it has become clear that the US and Japan are far ahead of the EU average and most Member States innovation performance as well (EC 2004a). This is clearly reflected in a composite indicator, the Summary Innovation Index (SII), which gives an overview of the aggregate innovation performance of a region. In the period 1996–2005, the average SII of the EU-15 was relatively constant at 0.40, but with the enlargement to EU-25 it went down to 0.34, while in the US it increased in the same period from 0.58 to 0.70, and in Japan from 0.60 to 0.72.

Two factors contribute to this widening of the innovation gap between the three economic blocs of the developed world. First the innovation performance of the business sector, and second, EU enlargement.

To a large extent, the gap between the US and the EU is due to three indicators that are dependent on the activity of the US business sector: (1) the working population with tertiary education (which explains 26 per cent of the gap); (2) patents (50 per cent of the gap); and (3) R&D expenditures, and mainly business R&D (11 per cent of the gap). All together they show that US businesses employ more educated personnel, perform more R&D, and patent more. The US business sector is more knowledge-intensive and more innovative. On the other side of the Atlantic, in Europe, larger companies rationalising their global development are not planning new research investment in the EU, but rather in other more attractive regions such as the US and south Asia. At the same time, European small and medium-sized companies find their ability to invest in R&D and innovation limited, by both the reduced self-financing capacity and the lack of internal R&D resources (European Commission 2004a).

The second factor is enlargement. Though it is not stated clearly in official EU documents, the enlargement from 15 to 25 and recently to 27 Member States was followed by a lowering of average EU innovation performance. This should be expected as most of the new Member States have less knowledge-intensive economies and major weaknesses in the successful structuring of innovation systems. Most of the EU regions, and especially those located in the periphery, show reduced R&D and innovation performance at all levels; spending, patents, and skills. The new European economy of 27 Member States is less innovative, and the EU has to cover a larger gap in order to reach the innovation levels of Japan and the USA.

A major drawback was also that the Lisbon Strategy failed to produce the expected results. Innovation is the cornerstone of this strategy, agreed by the Lisbon European Council in March 2000 and re-affirmed in Barcelona, March 2002. The strategy relies on abundant statistical evidence showing strong relationships between innovation and growth. Innovation is seen as the driving force behind increased productivity, competitiveness and high growth rates.

Within the context of this strategy, innovation management is focused on improving the environment for innovative companies, getting closer to markets and 'lead markets' especially, sustaining innovation in the public sector, and the regional dimension of innovation. Recommendations focus on R&D and innovation spending, which should be increased, stronger coordination between public and private-funded research, and the increased use of Structural Funds for R&D projects. Some specific European features are also considered, such as the large size of the public sector which should be heavily involved in the campaign to boost innovation; the regeneration of European cities which should orientate themselves towards the provision of knowledge, skills, and a qualified workforce; and European diversity and openness to innovative newcomers, including foreign-born individuals bringing new ideas and a spirit of enterprise (EC 2003). The creation of the European Research Area (ERA) was also a major component of the same strategy for making EU the most knowledge-driven economy in the world. ERA is expected to stimulate innovation, economic and employment growth through the exploitation of R&D results in areas such as biotechnology, ICT, nanotechnology, and clean energy technologies. Measures that enforce ERA include benchmarking of research policies, mapping of R&D excellence, increasing the mobility of researchers, strengthening of infrastructures, networking, boosting private investment, better intellectual property rights regulation, networking, and regional involvement (European Commission 2002a).

The quantitative scope of the Lisbon Strategy is to raise the average research level to 3 per cent of GDP by 2010, of which two-thirds should be funded by the private sector. The objectives sought (innovation, ERA, 3 per cent, and two-thirds) are expected to lead the European Union towards a stronger knowledge-based economy and bridge the growing gap in the levels of research and innovation between the US and Japan.

However, the mid-term evaluation of the progress made in 2004 showed that the outcomes of the Lisbon Strategy were somewhat disappointing:

> Taking stock five years after the launch of the *Lisbon Strategy*, the Commission finds the results to date somewhat disappointing and the European economy has failed to deliver the expected performance in terms of growth, productivity and employment. Job creation has slowed and there is still insufficient investment in research and development.
> (European Commission 2005)

This evaluation led to a revised strategy that was agreed in 2005. The new orientation did not change the original intention or the name of the strategy, but it focused on three key priorities: (1) attracting investments – making Europe a more attractive place to invest in and work; (2) more innovation – knowledge and innovation for growth; and (3) employment – creating more and better jobs. In addition, a new mode of governance was put in place, with a stronger national character and the obligation of the Member States to produce National Reform Programmes. The Structural Funds for the 2007–2013 period adopted a new strategic approach also, to ensure that the Funds sustain the Lisbon agenda for growth, jobs, and innovation. Investments related to the Lisbon agenda, in human capital and human resource development, business competitiveness, research and innovation, and the information society, should cover about two-thirds of investment through these Funds. Europe is far away from becoming the most knowledge-intensive economy in the world. At least the vision is still alive!

Innovation reshaping core and periphery

In the wave of developments taking us towards knowledge-based economies, cities and regions are transforming. The concepts of core and periphery are also changing. Today the core is what holds knowledge and technology. The periphery is what follows standardised forms of production, with low added value and complexity.

The new European core, for example, is comprised of regions of excellence which achieve best performance rates in key parameters of the knowledge-based society: education levels, R&D and knowledge generation, innovation, and the information society. Regions illustrating the new core include Uusimma in Finland, Stockholm in Sweden, Noord-Brabant and Zuid-Holland in the Netherlands, the south-east region of England including London, the regions of Île-de-France and Rhône-Alpes in France, Bayern and Baden-Württemberg in Germany. These are highly innovative regions, some of which are intensely globalised. The interesting thing is that in some cases their strong technological base is very new, coming out from a recent agrarian past.

The geographical position of the new European periphery is not necessarily the outer edges of the European area, in the south and north. Peripherality does not have the border geometrical layout of the less developed areas of the 1990–2000 period and the second Community Support Framework. Geographical disparities in the knowledge-based society are wider, more unbalanced, and extended. The regional technological gap measured by the R&D, patents, researchers, and other relevant indicators is much higher than the development gap, which is measured in terms of GDP and employment. Regional inequalities in terms of knowledge, both at EU level and at MS level, are expanding and changing form.

In the new setting of peripherality, innovation plays a very important role. This is captured in many empirical and statistical studies which show that among the key factors of regional development (workforce skills, investments, innovation, infrastructure, competitiveness), innovation is the most important one for improving productivity and wealth, both in manufacturing and the services sector (Hall and Hardy 2003).

The priority of innovation in regional development and shaping core-and-periphery was immediately reflected in regional policy. The efforts of the European regions to meet the challenges of an innovation-driven development led to a large number of Regional Innovation Strategies being elaborated and implemented. From 1995 – when these initiatives made their first appearance – to the present day, regional innovation strategies such as Regional Technology Plans (RPTs), Regional Innovation Strategies (RIS), and Regional Innovation and Technology Transfer Strategies (RITTS) have been implemented by more than 150 regions in all 27 EU countries. The aim of any RIS is to bolster the endogenous technological basis of that region and improve the ability of regional organisations to develop new products and technological innovations. A key concept underlying innovation strategies is that of a regional system of innovation as a network comprised of enterprises, research institutions, technological intermediation organisations, funding institutions, and technology consultants. Innovation springs from their synergy and systemic relationships, and is the result of interaction rather than the individual efforts of any one organisation. The geographical agglomeration generates positive conditions for integrating and reinforcing systemic relations, which are supported both by spatial proximity that makes cooperation easier and by institutions that bolster networking and cooperation. Within a regional system of innovation the individual organisations will secure technological resources, learn best practices, adopt values and standards of action, discover cooperative innovation methods and identify partnerships and networks. They will be able to monitor the rate of technological renewal and innovation because they find themselves in an environment favourable to research, knowledge and learning.

Innovation turning a new leaf for cities

Innovation-led development marks a new era for cities too. In their historical journey cities have functioned as locations for collective defence and protection against external threats, as centres of aristocracy and administration of large agricultural populations, and then in the industrial age as locations of industrial production and reproduction of the labour force. Today, new functionalities are added as cities are transforming in collective learning, knowledge and innovation centres.

As in the past, the power of cities today lies in the joint efforts of their populations, in cooperation. A new aspect of cooperation is that it now extends to knowledge and creativities allocated among the population and organisations in cities. The importance of cooperation is emphasised by any major explanation of innovation dynamics, and mainly social capital theory, which attributes innovation to the ability of organisations to collaborate and advance collective learning and knowledge sharing (Landabaso et al. 2007).

The new role of cities in the knowledge and innovation-led society is routed in the power of spatial agglomeration to create systems which synthesise knowledge and skills scattered across the population (Sassen 2003). Every resident of a city and every organisation located there (be it a business, foundation, research centre or university) is a nucleus of explicit and tacit knowledge. Their relationships of cooperation determine how information and knowledge channels are created, and technologies transferred and exchanged. Due to synergies and systemic relations, knowledge resources of the whole agglomeration are much greater than the sum of the individual sets of knowledge and skills.

If we carefully examine the knowledge and innovation system in large cities, we can see that it is fragmented. It is not a single, uniform system but consistsof separate clusters of dense cooperation networks, which correspond to different fields of knowledge, science and technology. Within each cluster, cooperation is enhanced by proximity, support institutions, and information and knowledge spillovers. Next to the clusters, universities and research centres create additional networks of knowledge and cooperation. The city assumes the form of a polycentric system of innovative clusters within wider networks of knowledge and technology linking the clusters with R&D institutions and technology intermediation organisations.

However, a city's present does not delete its past. Old forms of functionality and cooperation in terms of production, social services, and consumption remain ever present but are enriched by new forms of cooperation in knowledge and innovation. This results in extremely powerful knowledge agglomerations being created, with global networks and influences, innovation metropolises which ensure that innovation supply chains take advantage of global opportunities, and advanced forms of technology and production in core regions combine with skills to be found in less developed and lower labour cost regions.

Trading innovation with intelligent environments

Why though is this happening? Why is innovation depending from cities and regions? The answer lies in understanding that the route to innovation passes through environments that encourage the gathering and fertilisation of creativities and individual skills.

Converting scientific knowledge into new products and services, which is the core process of innovation, is only feasible under an environment rich in resources for research, experimentation, financing, and entrepreneurialism. Innovation is not produced by a linear process of using and applying scientific research. This is a quite false oversimplification. Research is a crucial component of innovation; but it becomes fertile by mobilising various capabilities within a wider system of knowledge, risk taking and entrepreneurialism. As systemic theories of innovation came to show, the emphasis has shifted from the innovation process to the innovation environment. The innovation system is exceptionally complex and radically unpredictable for disruptive innovations. And with the globalisation of production and exchange, it has also taken on global dimensions.

Given the complexity of the system that generates innovation and the fundamental conceptual feature of innovation on being unpredictable, it is reasonable to assert that we cannot plan for innovation per se, at least in its radical or disruptive forms. However, we can plan for the environment within which innovation blossoms. This basic assumption and substitution is to be found in most recent innovation strategies, which try improving the environment rather than the innovation process itself.

This strategy is, in effect, probabilistic. It considers innovation as a continuous, random process and attempts to organise its environment. In this environment some attempts (innovation designs) will succeed, while others will fail. In fact, failure is more probable than success. It is not feasible to predict with certainty the outcome of each attempt. What is important is to experiment within an environment that maximises the likelihood of success of a portfolio of innovation designs. Success is a matter of intensifying efforts with higher probability. The probabilistic approach comes out clearly from a recent survey, which shows that in the years to come, innovation will supplant cost reduction and mergers and acquisitions as a main competitive strategy of corporations. Ninety per cent of executives recently surveyed consider that the introduction of new products and services is crucial to profitable growth, and they aim to improve their innovation performance by 30 per cent in the next three years (Booz Allen Hamilton 2006).

The second step of this reasoning concerns the type of environment that may sustain contemporary innovation given the turn to global innovation networks and supply chains. The answer deliberatively is 'intelligent environments' that enable information retrieval from sources scattered around the world, getting technologies wherever these are available, and enabling global real time communication and exchange.

The central concern in trading innovation with intelligent environments is how such environments can be created and how to optimise the probability they offer for successful innovation projects.

Intelligent environments

A traditional route to creating intelligent environments is people- and institution-focused. It follows the concentration of intellectual and social capital in an area. Edvinsson (2005) cites Ragusa as a city with a high social intelligence capacity using international contacts, intelligent offices, and ambassadors to detect signals from the surrounding world and adapt its policy accordingly. He identifies three guiding factors in the making of intelligent Ragusa: information intelligence, being well-organised to relate to the external structural and human capital; governmental leadership for providing structural capital as a precondition for wealth creation; and community spirit or values for bonding human capital with different structural institutional capital for the larger common good of the city.

It is clear that this solution is not based so much on the concentration of intelligent/creative individuals as on the ability of a community and its enlightened leadership to establish institutions which promote cohesion, an outward-looking approach, and the utilisation of its population's intellectual capital. Pertinent here is the concept of 'embeddedness' as these driving factors do not operate in a vacuum of social relations, but in contexts with specific and adapted social capital. Social capital is a critical factor also, sustaining networks, norms and values that facilitate cooperation within groups of actors. All features of social capital encourage cooperation and collective action. As Landabaso points out:

> In general four main features of social capital can be distinguished: (a) Social capital is a market-based social exercise based on trust, reciprocity, shared norms and institutions; (b) Social capital can provide a relational infrastructure for collective action which facilitates cooperation within and among groups as well as enlarges a capacity for networking leading to mutual benefits; (c) Social capital can improve collective processes of learning and constitutes a key element of knowledge creation, diffusion and transfer – all processes critical for innovation and regional competitiveness; (d) Finally, social capital cements value-based networks stimulating successful regional clusters as well as regional innovation strategies and policies.
> (Landabaso 2007, p. 111)

At the other side of this people- and institution-focused route is the creation of intelligent environments using IT and artificial intelligence exclusively. It is based on applications that generate intelligent virtual environments, ambient intelligence, and embedded systems. From this perspective an intelligent environment 'is a space where ordinary human activities mix seamlessly with computation in a way that enhances the functions of both' (Intelligent Environment Lab 2007) or 'any space where ubiquitous technology informs the learning process in an unobtrusive, social or collaborative manner' (Winters et al. 2007).

Intelligent virtual environments, as dominant contemporary forms of intelligent environments, are web-based applications in various fields (information management, e-learning, e-commerce, e-government, e-promotion, e-tools, etc.) using artificial intelligence (AI) technologies. AI and advanced information technologies on the web enable functions quasi-similar to human intelligence to be performed, such as comparing, selecting, informing, and learning. Several AI technologies may be used to this end: intelligent agents, case-based reasoning, game theory, fuzzy logic. e-Commerce applications and digital marketplaces, for instance, may be supported by intelligent comparison shopping agents (selecting products and comparing prices), intelligent selling agents (contacting buying agents and negotiating prices), and other filtering and information collecting agents (Limthanmaphon et al. 2004). Intelligent virtual environments are produced by combining intelligent techniques and tools, embodied in autonomous agents, with graphical representations of environments. However, in this case, only virtual space comes into existence.

In more advanced forms, intelligent environments may be based on the linkages between the virtual and the physical worlds. Ambient intelligence applications can create environments where humans are surrounded by computing and networking technology unobtrusively embedded into their surroundings. Sensors and RFID may disappear into the urban environment and continuously feed virtual spaces with information. In these environments numerous networked embedded computing appliances offer functionalities in response to the occupant's presence and behaviour, as well as aiding the normal activities related to work, education, entertainment, security, privacy, and healthcare. People may interact with computers in the same way they interact with other people, via voice, gesture, and movement (Steventon and Wright 2006).

Combinations of these two trajectories, intellectual and social capital on the one hand, and digital spaces which incorporate AI applications on the other, open up a diverse number of paths to creating hybrid forms of intelligent environment merging human skills, cooperation-based institutions, and the power of computing. The spaces thus created are multidimensional, produced by an interdisciplinary convergence of many science and technology fields. The realisation of intelligent environments, notes the IE07 Conference, requires the convergence of different prominent disciplines: Information and Computer Science, Architecture, Material Engineering, Artificial Intelligence, Sociology and Design, as well as technical breakthroughs in key enabling technology fields, such as microelectronics, broadband and wireless communication, smart materials and intelligent agents (Intelligent Environments 07).

An example that well illustrates the convergence of social capital and virtual spaces in a creative and innovative way is the COINs. The acronym stands for 'Collaborative Innovation Networks'; a COIN is a cyberteam of motivated people with a collective vision cooperating over the Web, by sharing ideas, information, and work, with a view to achieving their vision (Gloor 2006).

COINs have three fundamental characteristics: they enable innovation through collaboration; they induce collaboration under ethical codes; and thanks to the capabilities of the Internet their members communicate directly. Gloor mentions a series of examples of such communities in the fields of IT outsourcing, sales force optimisation, R&D, mergers and acquisitions, learning, software and distributed product development, running political campaigns, and charities, which combine collaborative work, swarm creativity, and online networks and communication.

However, for us the primary case of convergence between human and artificial intelligence is to be found in intelligent cities and other forms of agglomeration of people working together, learning in cooperation, and using virtual spaces and digital tools to innovate.

Intelligent cities constitute a discrete category of intelligent environments generated by the agglomeration of creativities, smaller systems of innovation that operate within cities (technology districts, technology parks, innovation poles, innovative clusters), and digital networks and online services. Their added value lies in their ability to bring together three forms of intelligence: the human intelligence of the city's population, the collective intelligence of innovation support institutions, and the artificial intelligence of digital networks and online services.

The term (intelligent or smart cities) is used to characterise areas (cities, regions, districts in cities, clusters) where the local system of innovation is supported by and improved via digital networks and AI applications (Komninos 2002; Intelligent Community Forum 2007). By using ICTs the local innovation system acquires a greater depth and range, while its functions become more transparent and effective. The city gains innovation capability, which translates into competitiveness and prosperity. In this sense, two key factors in intelligent cities are:

- the local or regional innovation system which guides the development of knowledge and technologies to organisations in the area (businesses, universities, technology centres, incubators, etc.); and
- the digital information and knowledge management environment, which enhances the provision of information, communication, decision-making, technology transfer, and cooperation for innovation easier.

Local innovation systems comprise a physical space (the agglomeration of people, organisations, and infrastructure), and an institutional space based on a relationship of trust, communication, knowledge flow, untraded exchange, and partnership. They can be developed in many ways. The classic solution is clusters, groups of collaborating organisations located in a relatively small geographical unit. Within the cluster, research and innovation capacities improve due to the specialisation and cooperation of its members. Individual clusters combine within wider regional systems of innovation. Their specialisation is transferred from the level of cluster organisation to the cluster itself. Complementarity and synergies are expanded via multi-level networks that join various clusters together.

Thanks to the development of communication and information management technologies, these systems of innovation have acquired a new dimension – a digital one – which complements and enriches the two previous ones, the physical and institutional ones. The development of digital cities and digital innovation spaces added a new dimension to established local innovation systems. The new digital dimension offers additional capabilities for knowledge processing and exchange: faster, more direct communication, information storage, information processing, knowledge management, agent-based assessment, and so on. Key functions of the overall innovation cycle such as information provision, technology transfer, product prototyping, and partnership can now be performed in digital space and be located anywhere in the world. Thanks to the digital dimension, innovation networks and local innovation systems have gone global.

Thus, intelligent cities and regions form multi-level systems of innovation where the technological innovation mechanisms are deployed in physical, institutional, and digital space. Main building blocks of intelligent cities are clusters and other types of territorial systems of innovation complemented by digital spaces facilitating strategic intelligence, digital communication, networking, and collaboration. A key function of intelligent cities is to manage knowledge flows in all forms: the creation of new knowledge (research), monitoring knowledge flows (intelligence), disseminating current knowledge (technology transfer), cooperating in applying knowledge (innovation), developing new activities based on knowledge (incubation), and managing knowledge remotely (e-government). Intelligent cities express the need for a radical redesign of cities in the age of the global economy and the knowledge-based society.

Undoubtedly the Internet and digital technologies have offered the capabilities for intelligent cities to emerge. However, the true reason for their emergence is globalisation of the innovation cycle and the need for communication and coordination of innovation systems which have acquired a global reach and whose components are scattered around the world. Cities are being transformed into intelligent communities under the pressure for incessant innovation within global cooperation networks.

2 Regions of innovation excellence

Learning from the best

A new model of regional development

A series of surveys and studies about the actual regional development and the geographical distribution of innovation in Europe was recently published. Their common feature is that they examine the European regions comparatively and were prepared as accompanying research to the European innovation policies implemented through the R&D Framework Programmes, the Innovation Programme, and the Innovative Actions of the European Regional Development Fund. They have created an important bibliographical corpus informing us about emerging trends in European regions as they move towards a knowledge-based economy.

This bibliographical corpus contains three major components:

- *Regional policy periodic reports*, such as the Sixth Periodic Report on the Social and Economic Conditions of the European Regions (EC 1999), as well as the Second, Third, and Fourth Report on Economic and Social Cohesion (EC 2001a; 2004b; and 2007). These texts comment on research and technological innovation dynamics at the regional level and bring to the fore current trends concerning the geographical dimension of innovation in the EU. Their analyses address the concern about geographical concentrations of innovation in Europe and show a clear acceptance of the region as a key field for managing innovation.
- The *Community Innovation Surveys* (CIS) series. The CIS was carried out for the first time in 1992. CIS-2 took place in 1996 and CIS-3 in 2001. CIS-4 was conducted in 2005 covering all 25 Member States of the EU (Eurostat 2007). The results of the CISs were analysed at national level using a common methodology and a new dataset on innovation, as well as at regional level focusing on innovation disparities (D'Agostino 2000). Though there is no methodological continuity among CIS-1 and the rest of the series, these surveys are a source of primary data which can be read in many ways. This data is also presented in *Innovation Statistics* in Europe, a publication describing the practices and views of enterprises in terms of

innovation (EC 2001b). Taken in conjunction with regional data from Eurostat, the CIS documents highlight the relationships between innovation, productivity, changes in production, and competitiveness (EC 2001c).

- *The European Trend Chart Studies on Innovation*, which codify policy measures in the field of innovation taken by Member States on the one hand, while on the other hand presenting a series of new indicators for evaluating the performance of Member States and EU regions in the field of innovation. The European Innovation Scoreboards (EIS) in particular, published every year since 2001, have opened up new horizons for assessing the performance of states, regions, and industry sectors in the field of innovation and the progress of the EU towards the knowledge-based economy. The methodology for the EIS is based on the 'Oslo Manual', a model of innovation measurement proposed by OECD, the European Commission, and Eurostat.

These large-scale surveys and research bring to the surface some highly interesting information and insights about regional development in Europe. Some major conclusions concern the new regional innovation hierarchy in Europe, and the emerging model of regional innovation-led development.

First, it becomes clear that the regions of Europe are moving towards a **new model** of development, which draws its dynamism from technological innovation and the capability to convert R&D into products and services. At the core of this new model are the processes of research, knowledge management, and technological innovation. Innovation support institutions such as funding and company incubation, as well as the system of technology transfer, play major development roles, enabling business to innovate and compete effectively on global markets. In this model, innovation is leading on development and not vice versa. While the creation of new products and production technologies favour economic growth and employment, the reverse relationship does not always hold true; regional development does not necessarily feed innovation.

Within this new knowledge-intensive development model certain regions in the EU excel, thus setting standards and good practice models for all regions. **Regions of innovation excellence** may be defined with respect to the main factors that characterise the current state of development: education, research, innovation, digital infrastructure, and competitiveness. What is surprising is that the current regional excellence reverses established hierarchies. Regions and countries that were powerful until recently, now find themselves in the middle ranks rather than in the best places in the excellence scales. Less well-known regions are now at the top of the hierarchy. Regions of excellence are usually smaller in size and are not necessarily located in the geographical heart of Europe, in the highest accessible area. The regions of Finland and Sweden are performing best under strong institutional guidance, showing that neo-liberal principles of growth are not the only viable solution to technological and regional development policies.

A third conclusion is that the ***core of excellence*** can be identified in a multidimensional regional system of innovation, which brings together and connects capabilities and skills within industry clusters, institutional mechanisms of innovation, and digital spaces and e-services, enabling the global reach of knowledge, technologies, and markets. The development of innovations at company level is sustained by this regional innovation environment, external to the companies, from which networking, institutional support, and technologies become available.

This bibliographical body on innovation and regional development in Europe allows us to improve the conceptual models we use to interpret the innovation-based dynamic and its contribution to regional development. A new round of theoretical discussion has opened. Innovation and regional development cannot be interpreted exclusively by theories of flexible specialisation and learning regions, but rather by the combination of technological spillovers, institutional mechanisms for learning and innovation, and ICT based knowledge management. In regions of innovation excellence all these processes occur simultaneously. Regions with uni-dimensional characteristics, such as the flexible specialisation districts in central Italy, building their capability on technological spillovers only, are not to be found at the top of the new hierarchy of excellence. The lesson of contemporary regional excellence is about collaborative innovation and intelligence and its multiple dimensions: human, collective, and artificial.

The rise of the innovation economy in the EU

A new innovation economy is emerging in Europe. All recent measurements and statistics show that innovation, knowledge, and human skills are the main driving forces of contemporary competitiveness and development. The turn towards knowledge and innovation, towards new products that integrate results of scientific discovery, is bolstered by the dominant technological paradigm today; a paradigm fuelled by a combination of information technologies, computers, telecommunications, broadband services, and the Internet. These horizontal technologies are progressively incorporated into every field of economic activity and industrial sector and they exhibit some distinctive characteristics from technologies that fed previous cycles of technological and industrial change. The new tools do not relate so much to the transformation of materials and manufacturing processes as to human intellectual capacity itself: they multiply our ability to communicate and our conceptual skills to resolve problems. Furthermore, they make it possible for every employee, even in the smallest company or in the most remote region, to make use of advanced analysis, classification, communication, storage, forecasting, and other advanced problem-solving methods. With respect to previous technological cycles, actual information and communication technologies amplify intellectual capacities, cooperation capabilities, and human intelligence.

24 *Globalisation of innovation and intelligent cities*

As the new development model steps forward, regions, enterprises and industrial sectors are looking for knowledge resources, wherever these are available, in order to obtain know-how and to build their competitive advantage. This is one basic feature of the new knowledge-based economy. The second, and equally important, feature is that development is increasingly structured around regional rather than national entities. The interplay between the regional and the global takes the lead. Knowledge-based development, regional innovation, and global supply chains go hand-in-hand.

The 2001 European Innovation Scoreboard (EIS) offered the first systematic account of the progress of Europe in the new innovation-driven economy. The Scoreboard was requested by the Lisbon European Council in March 2000, where a plea was made to bolster innovation within the European Union as a response to the challenges of globalisation and new competition. When it was published it provoked a great deal of interest about the new phenomena it had encapsulated in numbers, such as the leading position of Scandinavian countries in innovation performance, the disassociation of innovation performance from the size of a country, the progress achieved by the cohesion countries, etc. The 2001 Scoreboard contained comparative data on the performance of EU Member States across 17 indicators falling into four major categories: human resources, knowledge generation, new knowledge transmission and application, and innovation funding and markets. Gradually, the EIS was enriched with new indicators, 19 in 2003, and 26 in 2005, in order to capture more dimensions of the innovation-driven economy (EIS 2001–2006). The number of categories was also increased from four to five. The EIS-2006 contains the updated set of 26 indicators organised in five blocks:

- Five indicators are included in the **Innovation Drivers** block. They relate to science and technology graduates, the tertiary education of the population, lifelong learning, youth employment, and broadband penetration. They measure the structural conditions required for innovation.
- Five indicators are included in the **Knowledge Creation** block. They measure investments in R&D, which are considered a key element for successful knowledge-based development. In particular, they concern public expenditure on R&D, business expenditure on R&D, the share of medium-high and high-tech R&D, the share of enterprises receiving public funding for innovation, and university R&D expenditures financed by the business sector.
- Six indicators are included in the **Innovation and Entrepreneurship** block: SMEs innovating in-house, innovative SMEs cooperating with others, innovation expenditure with respect to turnover, early-stage venture capital, ICT expenditure, and SMEs using organisational innovation. These indicators measure the efforts for innovation at the company level, and the diffusion/absorption capacity of companies.
- Five indicators included in the **Applications** block measure output performance, expressed in terms of labour and business activities, and their

added value in innovative sectors: employment in high-tech services, high-tech exports, share of new-to-market products, share of new-to-firm products, and employment in medium- and high-tech manufacturing.
- Then, five indicators in the **Intellectual Property** block measure output in terms of successful know-how: patent applications to the European Patent Office (EPO), applications to the US Patent Office (USPTO), triad patents, community trademarks, and community designs.

The Innovation Scoreboard surveys have adopted a benchmarking approach for each indicator, as well as for composite indicators like the Summary Innovation Index calculated from individual ones. The Scoreboards show performance levels and deviations by Member States from the EU average, on the one hand, and on the other hand performance levels and deviations by Member States from the average change over the last five years. Many indicators also compare the performance of the US and Japan. In this way the benchmarking relates not only to the EU Member States and regions, but also to its two main competitors in the global economy.

Regional data were also made available in the EIS-2002, EIS-2003, and EIS-2006. Sectoral data by industry were published in the EIS-2003 and EIS-2005. Fewer indicators were included in the regional and sectoral scoreboards, but they enable us to identify regional and sectoral variations of innovation performance within the Union. A systemic approach was also introduced, looking at innovation input and output, supply and demand, innovation governance, and cluster analyses.

These measurements and data established a robust reference framework for comparing and evaluating innovation performance levels of the Member States, regions, and industries within the Union. At the same time, they document a series of interesting trends about the Union's position in the emerging global knowledge-based economy.

At the level of EU Member States innovation performances are very unequal. A division of the EU into four areas is clearly identified: Sweden, Finland, Denmark, and Germany are the leaders within the Union; the UK, Belgium, the Netherlands, France, Ireland, and Austria follow; Italy, Slovakia, Estonia, Spain, Malta, and Hungary are trailing behind them; and the Czech Republic, Cyprus, Portugal, Greece, Poland, Bulgaria, and Romania make the most backward, catching-up group (Figure 2.1). With the recent enlargements (EU-25 and EU-27) performances in all fields deteriorated, with the exception of Youth Education and High-tech Exports.

The performance of the Member States was further analysed with respect to the third Community Innovation Survey, which reveals significant differences in the innovation behaviour of the EU Member States. Based on this data, Abramosky *et al.* (2004) examined country-specific and sector-specific differences in innovation performance. For example, country-specific features include differences in the macroeconomic environment, competition, structures of factor markets, regulation, and technology policy including public support. On the

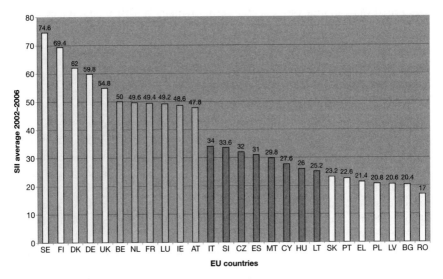

Figure 2.1 Innovation performance in EU Member States
Source: Based on EIS 2002–2006 data

other hand, the differences in innovation behaviour could also be due to varying sector compositions of economic activity given that innovation behaviour is sector-specific. The results reveal that the country-related – rather than the sectoral – differences have a greater role to play in explaining the differences in innovative performance with the exception of the innovation intensity indicator. With few exceptions (Luxembourg, Czech Republic) innovation performance of Member States in the period 2002–2006 remained constant (Figure 2.2).

In comparison to the US and Japan, the latter outperforms the EU in all indicators, except Community Trademarks and Designs. The US outperforms the EU in 11 out of 15 available indicators. However, in terms of global innovation performance two small EU countries, Finland and Sweden, hold first and second positions (Figure 2.3). The fact that in various indicators (i.e. global innovation index, public and business R&D, etc.) some Member States (Finland and Sweden) perform better than the US and Japan does not mean that the EU has a lead, even in these areas, because data for the US and Japan are aggregated for all their regions. Overall, the European Union suffers from two key weaknesses in patents and business R&D. The rapid increase in business R&D noted in Japan and the US since 1984 has widened the gap between these countries and the EU.

However, the fact that the first places in the world classification of innovation belong in Finland and Sweden, small countries with strong state intervention and institutional organisation, questions a strong neoliberal argument for the higher efficiency of market-mediated relationships in the innovation economy. State-led and market-led economies can equally well achieve very good and very bad records of innovation. It is more a question of organisation and social capital than primacy of market or state forces. For small countries that do not have big multinational companies, a good level of institutional intervention is necessary to counterbalance the advantages that the multinational companies acquire from the globalisation of innovation networks.

Data at the regional level document very clearly that innovation is a strong driver of regional development. GDP per capita (x) is positively related to the Revealed Regional Summary Innovation Index (RRSII) (y). Their relationship is: y= 0.0655x + 0.573, and R2= 0.0556 (EIS 2003). However, in southern European regions (Greece, Portugal, Spain) the relationship is flatter, and increases in RRSII lead to comparatively smaller increases of GDP.

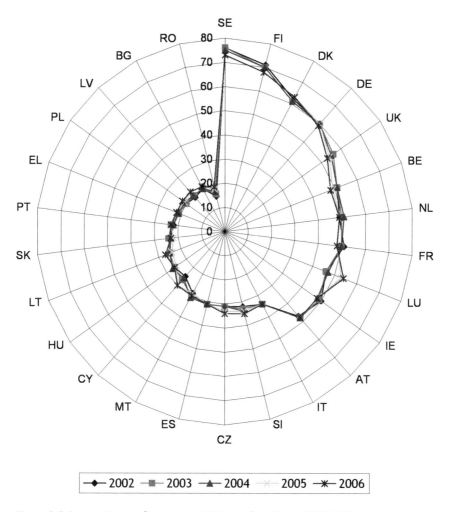

Figure 2.2 Innovation performance in EU Member States 2002–2006
Source: Based on EIS 2002–2006 data

28 *Globalisation of innovation and intelligent cities*

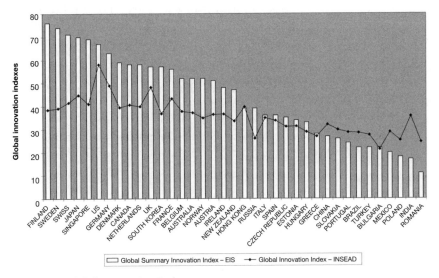

Figure 2.3 Global Innovation Index
Source: Based on EIS and World Business data

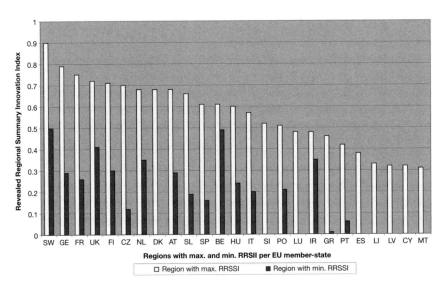

Figure 2.4 Regional disparities in innovation performance
Source: Based on EIS 2006 data

In the interior of Member States regional innovation disparities are more important than between the Member States (Figure 2.4). Intra-regional innovation policy at the level of EU Member States is likely to bring more significant results than inter-regional EU policy.

At EU business level data documenting the primary role of innovation in business development are given by the European Competitiveness Reports (EC 2001c and 2004). The 2001 report examines, inter alia, the statistical correlation between four categories of indicators (research, human resources, IT and communication technologies, innovation) on the one hand, and on the other hand the increase in manufacturing production and productivity. The indicators used are similar to those of the Innovation Scoreboard. Some conclusions that should be highlighted are:

- Increases in production and productivity in manufacturing are positively associated with inflows of research, patents, and the number of publications per resident. Although these correlations are not strong, those between the increases in production and publications and the increases in productivity and patents are statistically significant.
- None of the factors relating to human resources present any significant correlation with increase in productivity and only public expenditure on training and the working population with university level knowledge show a correlation with the increase in production.
- From the indicators relating to IT and communication technologies, Internet dissemination and the number of personal computers per resident showed a positive correlation with the increase in production and productivity.
- Innovation indicators (expenditure on innovation, technological collaboration, ongoing research and development) show the strongest correlation with the increase in production, a fact which is interpreted by the Report as confirmation of the evolutionary theories of innovation and development. On the contrary, the participation of new products in sales is not statistically significant for the increase of production and productivity.

Data from the EIS, the Community Innovation Surveys, and Competitiveness Reports document the primary role of innovation in European development at the level of Member States, regions, and enterprises. At any level, the innovation economy is described using the same type of indicators, stemming from the Oslo Innovation Manual:[1] inputs in terms of research, human capital, innovation capacity, innovation, and information technology use; outputs, in terms of wealth generation, production, and productivity. This documented turn towards an innovation-driven economy in the EU received official political recognition and sanction at the Lisbon Summit, March 2000. The Summit endorsed the commitment of the EU towards innovation, a target (Lisbon) that is now actually shaping all European policies.

Regions of innovation excellence in Europe

It is widely known that the geographical distribution of research, knowledge, and technological innovation is particularly uneven. A small number of cities and regions in Europe and the US account for the majority of researchers, laboratories, patents, and research and technological innovation resources. In the less-developed regions the resources for research and technology are rarer and performance levels lower. Nonetheless, this regional technological gap is not a specific European feature. It also occurs in the US where geographical research documents that major technological complexes have been developed in California, New York, Massachusetts, and North Carolina, (John Adams Innovation Institute 2007; Saxenian 1990; Scott 1988a); and in Japan as well, where the MITI 'Technopolis' programme attempted to overturn the large concentration of research and technology in the Tokyo-Osaka region (Masser 1990; Tatsumo 1986).

Discussing the concentration of R&D and innovation in a few regions globally, the Index of Silicon Valley (2007) introduced the concept of 'spiked world' meaning that although the global competitive field is becoming flatter, regions still vary by their relative strengths and weaknesses from which regional specialisations and comparative advantages emerge, creating 'spikes' over a flat world. The identification of 'spikes' was based on the rankings of three critical factors affecting regional systems of innovation: (1) employment in information technology per capita, (2) patents per capita, and (3) venture capital per capita. A number of global innovative regions were identified including Silicon Valley, Seattle, Austin, Raleigh, and Boston in the US; Stockholm, Munich, and Helsinki in Europe; Israel in the Middle East; Bangalore, Beijing, Seoul, Shanghai, Taiwan, and Tokyo in Asia. The challenge for each region is to recognise its own strengths, identify regional 'spikes' based on defined advantages, and then connect to other global 'spikes' for mutual benefit.

In Europe a series of studies have identified such 'spikes' of innovation excellence and 'gaps' as well, and discussed their trajectories of development. The first attempt to identify regions of excellence, in terms of research and technological innovation, goes back to 1992, in the FAST study on the 'Islands of Innovation' in Europe (EU-12), which is also known as the 'Archipelago Europe' study.

'Archipelago Europe' ranked the EU regions in relation to their level of employment in the R&D sector. It did not include the Scandinavian countries since they had not yet acceded to the EU. The islands of innovation identified were London, Rotterdam, Amsterdam, Île-de-France, the Ruhr, Frankfurt, Stuttgart, Munich, Lyon, Grenoble, Turin, and Milan. In parallel with employment, these areas also accounted for 75 per cent of public R&D financing. A dense network of enterprises and research centres created a web of activities which sustains a continuous generation of new products and other innovations (EC 1994).

Later, with the publication of the second Report on Economic and Social Cohesion (EC 2001a), a new set of regional data came to the surface, informing both on innovation input and output factors. The report presented comparative data on the 'level of education of the working population', which characterise the available human capital in a region, 'patents applications', and 'GDP per capita' which is a usual indicator of regional wealth. With respect to the above three indicators the Report defined the hierarchy of European regions in terms of innovation inputs and generated output. In Table 2.1, EU regions with higher than average performance in all three indicators were classified with respect to the number of indicators they excelled in. This ranking and characterisation of regional excellence is different from the approach taken from the older 'Archipelago Europe' study because it takes into account not only performance levels in terms of knowledge and innovation, but also considers performance levels based on regional wealth.

More recent comparative analysis of regional innovation performance is to be found in the 2006 European Regional Innovation Scoreboard (Hollanders 2006), which is an update of two Regional Innovation Scoreboards that were published in 2002 and 2003 under the European Commission's 'European Trend Chart on Innovation' (EIS 2001–2006).

Table 2.1 Regions of innovation excellence in the EU

	Regions	Per capita GDP (EU-15=100, 1998)	Patents per million of population (average 1997–1998-1999)	High level of education (% of population 25–59 years old, 1999)
1	Uusimaa (F)	141.5	355.3	41
2	Stockholm (S)	136.1	464.9	39
3	Region Brussels (BE)	168.8	134.5	36
4	Oberbayern (G)	161.2	571.0	29
5	Île de France (F)	151.7	252.7	32
6	Berkshire, Bucks, Oxfordshire (UK)	130.2	227.0	37
7	Karlsruhe (G)	130.0	496.7	28
8	Stuttgart (G)	122.5	416.4	25
9	Noord-Brabant (NE)	111.9	445.4	21
10	Chesire (UK)	111.6	167.0	32
11	Utrecht (NE)	142.4	139.5	30
12	Zuid-Holland (NE)	131.9	121.5	26
13	Rhône-Alpes (F)	100.8	202.3	24
	EU-15	**100.0**	**119.4**	**21**

* Regions are classified with respect to the number of indicators in which they excel.

Source: Based on data from the 2nd Report on Economic and Social Cohesion, EC (2001a)

The 2002 and 2006 scoreboards were based on the same set of indicators: seven innovation-driven indicators for which data were available, but for the EU-15 (2002) and the EU-25 (2006). These indicators are (1) Human resources in science and technology; (2) Participation in life-long learning; (3) Employment in medium-high and high-tech manufacturing; (4) Employment in high-tech services; (5) Public R&D expenditures; (6) Business R&D expenditures; and (7) EPO patent applications. The 2003 regional innovation scoreboard was based on 13 indicators for the EU-15 Member States.

Based on this data, the Revealed Regional Summary Innovation Index (RRSII) was calculated for 173 regions in the EU-15, which identified the most innovative regions with respect to the national and European performance levels.

The ten most innovative EU regions identified in 2002 were Stockholm (SE), Uusimaa (FI), Noord-Brabant (NL), eastern region (UK), Pohjois-Suomi (FI), Île-de-France (FR), Bayern (DE), southeast region (UK), Comunidad de Madrid (ES), and Baden Württemberg (DE). The inclusion of Madrid in the ten best EU regions shows that regions from Objective 1 countries can hold leading positions on the innovation scale. A positive relationship between the RRSII and regional GDP was also documented, which indicates the linkage between innovation and regional development. Nonetheless, regions with very high living standards are not included among the top ten; many other factors can generate high incomes too. On the other hand, the case of the Noord-Brabant region having a modest GDP per capita shows that a powerful innovation performance does not always lead to high incomes.

In 2006, the top ten performing regions identified among the 208 regions of the EU-25 were Stockholm and Västsverige in Sweden, Oberbayern (DE), Etelä-Suomi (FI), Karlsruhe (DE), Stuttgart (DE), Braunschweig (DE), Sydsverige (SE), Île de France (FR) and Östra Mellansverige (SE). Their performances reflect national innovation strengths, with the majority of regions coming from the group of innovation leading EU Member States (SE, FI, and DE).

The differences in the four rankings above (Archipelago Europe, second Cohesion Report, Innovation Scoreboards 2002 and 2006) may be explained by the differences in the indicators used in each case, and the way in which the indicators are combined together. The simplest method is used is Archipelago Europe, which employs only one indicator, the level of employment in research and technology. The 2002 and 2006 Innovation Scoreboards use seven indicators, from which they construct a composite one (RRSII). However, in some cases the RRSII gives a misleading estimation of regional innovation performance because of the way the seven individual indicators are aggregated. The indicators from the Cohesion Report were not aggregated into a composite index; the regions were classified with respect to the number of indicators they excel in.

The aforementioned assessments of regional performance in the knowledge and innovation sectors, despite the differences highlighted, reveal a core of regions of innovation excellence around which the four evaluations seem to converge. Taking into account the absence of Scandinavian regions from the Archipelago Europe analysis, the assessments mainly differ in relation to the position of the regions of northern Italy (Lombardy and Piedmont), whose performance in patents and educational level indicators is not further confirmed. Taking all the assessments together, we may argue that the top regions in the EU in terms of technological innovation are:

- Two Scandinavian regions: Uusimaa in Finland, and Stockholm in Sweden;
- Two regions in Holland: Noord-Brabant, and Zuid-Holland;
- The southeast region including London in the UK;
- Two regions in France: Île-de-France, and Rhône-Alpes; and
- Two regions in Germany: Bayern, and Baden Württemberg.

The core of excellence: regional systems of innovation

What distinguishes the regions of innovation excellence from other regions is their capability to bolster the innovation performance of the organisations which have established themselves there. Whether we are referring to an enterprise, a research laboratory or a technology provider, locating in a region of innovation excellence gives additional research and innovation capability. This additional strength, which derives from the territory, its people and resources, is the reason that pushes to the selective location of R&D departments and innovative enterprises in the most technologically advanced regions. The vehicle for regional innovation excellence is the endogenous capability of cities and regions to advance technological innovation, based on a combination of skills, institutions, and infrastructures. Thus, the core of regional innovation excellence lies in the systemic relationships which bridge knowledge, skills, and funding offered to the organisations located there.

How can this be done though? How is the core of excellence established? In order to illustrate the mechanism that enhances innovation within a wider territorial system, we will present a simple model of technological innovation.

Innovation developed by an enterprise is about new products and services, new production technologies, and new organisational processes. As the Community Innovation Surveys brought to the fore, the priorities of EU enterprises in product, process and organisational innovations are well defined. For example, in developing new products, priorities are focused on (1) improved quality; (2) access to new markets; (3) product extensions; and (4) old product replacement. In production technologies, priorities are focused on (1) reducing the cost of labour; (2) advancing flexible production; (3) reducing material waste; and (4) reducing energy consumption. Organisational innovations are associated with adapting to standards and environmental protection regulations (EC 2001b).

In most enterprises, achieving these goals is based on the operation of internal research and development departments, which experiment on new products and processes, exploiting knowledge obtained from developments in science, technology, and the marketplace. Where required, skills and technologies are transferred from external providers via licensing or other technology transfer agreements. External contract research, in delimited areas, is also assigned. This is a usual innovation model, based on the internal capacity of enterprises vis-à-vis research, financing, and new product development. Product development goes through a series of interconnected stages which, step-by-step, address R&D, market research, new concept development, production re-tooling, and production (Figure 2.5).

The in-house technological innovation model is essentially linear. Innovation starts from the R&D department, and following the construction of prototypes, the production and marketing departments become involved. In newer versions of the process, linearity is limited to the operation of planning teams in which scientists from the R&D department participate together with production engineering and quality-control experts and executives from the marketing department (Iansiti 1993). However, even this more rounded approach does not negate the linear and closed character of the in-house innovation generation process.

In regions of innovation excellence this model is radically overturned. The presence of significant resources and technological capabilities located outside yet in the vicinity of enterprises permits a radical restructuring of the linear, in-house model. The space within which innovation is developed expands from the inner world of enterprises into the regional or inter-regional space. Research results from universities and R&D institutes in the area are utilised; financing from venture capital or other sources is offered; new product development services are provided by technology consultants; expertise and skills become available from specialised technology transfer centres.

The change in the space within which innovation occurs is accompanied by a change in the type and character of activities involved in the innovation

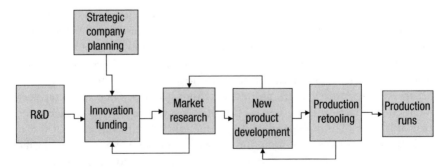

Figure 2.5 In-house innovation

process. New activities take part in a continuous cycle of innovation, which now takes the form of networks linking collaborating organisations. Within these networks the linearity is overturned. Innovation may initiate from any element of the networked system, from technology transfer centres, funding organisations, new product development specialists, and consultants. An additional blow to linearity is due to the change of the financing conditions, which now do not depend on company funds exclusively, with many alternative options offered from venture capital, business angels, regional incentives or even other enterprises that open up. The environment becomes much more complex and the opportunities and potential for innovation multiply, sometimes due to good partnership between enterprises, sometimes due to state aid, and sometimes due to promising public research activity.

The overall process takes the form shown in Figure 2.6, which illustrates the externalisation of new product development processes from the enterprise to the region, and the addition of a second layer of innovation processes over the company: the regional system of innovation. Consequently, what characterises the regions of innovation excellence is the operation of this second layer of innovation, creating an external system in which five basic innovation functions take place and interconnect: research, funding, technology transfer, new product development, and innovation supply networking.

In the recent literature on the development of innovative regions in Western Europe and the US, the contribution of the external environment that promotes developmental and innovation excellence in certain territories was attributed to a **spatial competitive advantage** generated due to agglomeration, specialisation, partnership relationships, and institutional learning. Two theories that formulated the milestones of this interpretative framework were the 'industrial district' theory (including cluster theory) and the 'learning region' theory.

At the end of the 1970s Italian geographers and economists (Bagnasco 1977; Becattini 1979; Brusco 1982) laid the foundations of a new theoretical paradigm which interpreted the growth of central regions of Italy with respect to system-areas and industrial districts. The interpretation was rapidly transferred to the other side of the Atlantic by Piore and Sabel (1984); in their influential 'second industrial divide' they treated the industrial district as the spatial model for flexible accumulation at the end of the twentieth century. To some extent the same theory was applied by Saxenian (1990); Scott (1988a and 1988b); and Storper and Scott (1988) to describe new industrial spaces and technopoles in California and Massachusetts, based on information technologies and the computer industry.

Becattini (1989) explained that the spatial competitive advantage emerging in the industrial district was due to collective creativity and cooperation among specialised skills within the district, as the district functions as a creative environment. On the contrary, the explanation given by Scott emphasised the reduction of transaction costs and the positive external economies created from spatial proximity within the district. The 'District' theory and the systemic results of

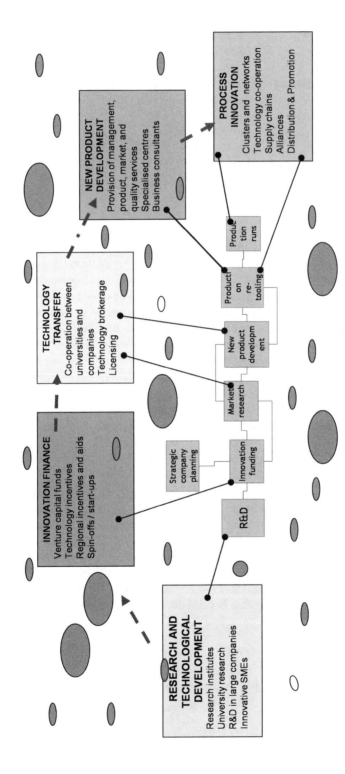

Figure 2.6 Innovation within the regional system of innovation

collaboration between enterprises were enriched with the work of Porter (1990) and Krugman (1991) on clusters. As Wood (1991) has accurately noted:

> industrial clusters, in their much talked about form, were discovered by Alfred Weber (1929) 100 years ago and have been revived quite a few times since then by Marshall (1919), Hoover (1948), Isard (1951, 1956) and more recently by Scott (1990), Storper (1993, 1997), Porter (1990) and Krugman (1991).

Later, in the 1990s a new version of the spatial competitive advantage was formulated, to some extent as a continuation of the 'district' theory. The most valuable contributions by Cooke and Morgan (1998); Maskell and Malmberg (1995); and Storper (1997) dealt with technologically advanced regions as spatial entities having capacities for learning, innovation, and adaptation. The 'learning region' theory preserved many of the features of flexible specialisation and in particular networking and partnership between enterprises, but added a new important element that was institutional and organisational in character. The competitive advantage of a region operating as an integrated system was attributed by Storper (1997) to un-traded interdependencies, while by Cooke, Uranga-Gomez, and Extebarria (1997 and 1998) to institutional research and technological development mechanisms, innovation financing, and specialised consultancy. The impact of evolutionary economic thinking and in particular the ideas of Lundvall (1992), and Nelson and Winters (1982) were also acknowledged by most writers.

'Industrial districts', 'clusters' and 'learning regions' today provide the central interpretive schema for regional innovation excellence. The core of excellence lies in generating a spatial competitive advantage; however, there is no agreement about the nature of this advantage. Our version is that the latter is determined by the most central feature of the knowledge-based economy, the ability to set knowledge networks that constantly turn scientific research into new products and production technologies (Kafkalas and Komninos 1999).

Regions of technological excellence are not those which have the most R&D laboratories, the most patents, the most technologically advanced sectors, or the most innovative enterprises. Quantitative superiority is the result not the cause. Regions of excellence are not simply the sum of their innovative organisations. On the contrary, they are structures for generating innovation. **Above all they are systems**. Critical indicators for identifying them relate to the presence of organisations that guarantee the conversion of technological knowledge into products (venture capital, incubators, technology parks, spin-offs, consultants specialised in product development, etc.). Critical indicators are also those which trace systemic relationships, such as innovative enterprises which do not have their own R&D departments, products generated with the involvement of new product development centres, technology transfer agreements, knowledge and technology collaboration networks, etc.

The ability to generate innovation has now become so demanding and complex that the entire social and institutional system in an area needs to be re-organised in order to ensure the necessary capacity of organisations in developing innovations. Cities and regions are forced to comply with this logic.

Making the core of excellence: four stages of evolution

Surveys and monographs on regions of innovation excellence in Europe (Cohen et al. 1997; Moreno et al. 2005; Saperstein and Rouach 2002; Simmie 2001) show their extreme diversity and path-dependent trajectory. The territories of innovation differ substantially in terms of size, content, activities, and profile; but, above all, and this is of interest to us, they differ radically in relation to the way the respective innovation system is structured and operates within each region.

Characterising regions with respect to their system of innovation is an extremely difficult task, mainly because of two overlapping geographies. The geographical boundaries of administrative regions are not the same as the boundaries of regional innovation systems. For instance, the boundaries of regions into which the EU has been divided for administrative/planning purposes (NUTS 2 planning regions) do not coincide with the boundaries of functional EU regional systems of innovation. The latter may consist of organisations located in some areas of a planning region or in different planning regions. It is possible for two or more innovation systems to operate within the same planning region; vice-versa, the same innovation system may extend into more neighbouring regions. This problem does not arise so much in demarcating industrial districts and clusters, which constitute rather small geographically concentrated systems of innovation but for wider systems composed of hundreds of companies, R&D and intermediary organisations. The different geographies of planning regions and regional systems of innovation were one of the main focuses of criticism against the learning region theory (Lovering 1999; McLeod 2001), but this does not affect the concepts and intellectual constructs of the theory; rather its political dimension and utility in policy-making. This is especially relevant for regional innovation strategies exercised at the level of administrative and planning regions, while the underlying system of innovation is inter- or intra-regional.

While it seems complicated to characterise planning and administrative regions with respect to integral systems of innovation, it is feasible to think about the geography, typology, and genesis of territorial innovation systems. For instance, typologies of innovation systems were attempted in terms of conceptual categories (Markusen 1996), demographical characteristics (density, range, activity, development prospects, innovation capacity, industrial organisation, governance mechanisms) (Enright 2000), and relationships among innovation factors in various regions (D'Agostino 2000).

Attributing the making of innovative regions down to the external system sustaining the ability of organisations to innovate and constantly improve products, services, and processes, the immediate question that arises is how can this system be created? In what way have the best regions in Europe managed to establish systems which bolster innovation and the continuous conversion of knowledge into products?

We will approach this question from an evolutionary point of view, which brings to the surface successive stages of systemic relationships creating the overall innovation system that operates into a region or territory. The innovative character of each region is defined by its progress on a ladder of development towards integrated and multi-level innovation systems from which enterprises obtain information, knowledge, skills and resources to innovate. Four stages are important in this evolution.

1 **Absence.** The starting point is the absence of spatial competitive advantage based on a regional system of innovation. Enterprises may generate new products and production technologies, but rely on their own internal capabilities for designing, funding, and developing innovations. The regional environment does not contribute to their innovation capability. This does not mean that the region does not show some level of innovation performance. The typical innovation indicators (public and private expenditure on R&D, patents, R&D employees, high-tech sectors) may have a low, average or high value, but this depends exclusively on the ability of enterprises and not the region.

This layout is normally encountered in most European Union Objective 1 regions where traditional industrial sectors dominate the production landscape, research is primarily carried out by universities, and the local web of technology transfer, financing and intermediation is little developed. It may also be the case of innovative regions where large multinational enterprises are located whose technological capabilities greatly exceed regional resources. There are indications that the Noord-Brabant region in the Netherlands, for example, which we ranked among the top ten innovative regions in the EU, owes its performance more to the presence of Phillips than to a regional innovation environment. It is a medium-sized region with 2.4 million residents having the fourth best performance level in the EU in terms of patents. At the same time it is a region with a strong agricultural character and a population educational level at EU average. This structure would not have led it to the peak of regional excellence in intellectual property without the presence of Phillips, whose activity is associated with Eindhoven, a city closely identified with the company, which was established there and continues to have its headquarters there.

2 **Clusters.** A precarious form of the local innovation system appears within the separate industrial districts/clusters of the region. Companies develop local networks (horizontal or vertical) and obtain knowledge and innovation capacity from associated enterprises and organisations.

In Stuttgart, for example, which is a major centre for technological innovation in the heart of Baden Württemberg, Southwest Germany, three different production clusters co-exist: (1) machinery manufacture; (2) engineering; and (3) electronics. The main competitive advantage of this city of 2.6 million residents is generated by the specialisation in knowledge-intensive industries across these clusters. Three-quarters of people employed in manufacturing work there, accounting for 40 per cent of all industrial enterprises in the region (Strambach 2001). The region has the highest degree of people employed in manufacturing in all German metropolises and the lowest level of tertiary sector employment. The clusters are based on different areas of science and

Table 2.2 Dimensions and typology of regional clusters

Dimension	Types	Cases
Geographic scope	• Localised • Dispersed	Sassuolo ceramic tiles Japan synthetic fabrics
Density	• Dense • Sparse	New York financial services New Hampshire instrumentation
Breadth	• Broad • Narrow	Osaka electronics Dalton tufted carpets
Depth	• Deep • Shallow	Danish agro-industry Ireland pharmaceuticals
Activity base	• Activity-rich • Activity-poor	Silicon Valley Chihuahua maquilas
Growth potential industry growth/ competitive position	• Sunrise/competitive • Sunrise/uncompetitive • Noonday/competitive • Noonday/uncompetitive • Sunset/competitive • Sunset/uncompetitive	Los Angeles multimedia Quebec transport equipment Boston minicomputers
Innovative capacity	• High innovation • Low innovation	Boston biotechnology, Milan fashion Singapore electronics
Industrial organisation	• Core-ring with coordinating firm • Core-ring with lead firm • All ring, no core • All core, no ring	Veneto garments Toulouse aircraft Carpi knitwear
Coordinating mechanisms	• Spot markets • Short-term coalitions • Long-term relationships • Hierarchies	Prato textiles Hollywood motion pictures Turin factory automation equipment Detroit autos

Source: Based on Enright (2000)

technology, having separate cooperation networks and forms of collaboration. Enterprises outside of these districts enjoy a much poorer innovation environment.

Industrial districts and clusters operate in all regions of excellence that we highlighted. Usually they do not cover the entire region, but only parts of it. The type of districts and the small innovation systems created inside them vary considerably. The extensive survey of Enright (2000) on clusters in US, Europe, and Asia shows this variety. Different types of clusters are defined with respect to their fundamental dimensions and characteristics (Table 2.2).

Markusen (1996) proposed another typology of districts distinguishing four types:

- marshallian industrial districts comprised of a dense network of small local enterprises;
- hub-and-spoke districts, comprised of a few large local enterprises;
- satellite-platform districts, such as a cluster of branch plants which belong to large enterprises outside the area, usually multinationals; and
- state-focused districts, where a public authority (university, R&D laboratory, public sector organisation) is the main owner and organiser of the cluster.

In each type of district/cluster, the enterprise collaboration mechanism, the innovation generation mechanism, and the type of resources that the district provides differ substantially.

3 **Multi-cluster regional systems.** Industrial districts and islands of research and technology tend to come together in a wider regional system of institutions and partnership networks. This environment offers research outputs and resources from universities and research centres, financing from regional incentives or venture capital funds, technological partnership opportunities for smaller enterprises, ongoing training, and new product development services. A resource-rich innovation environment characterises the majority of regions of excellence we highlighted. Numerous clusters operate inside the regional innovation system, which link individual enterprises to research, technology learning, and financing organisations. The role of institutions, the governance system, regional administration, local business groupings, and other institutional-type initiatives are important in generating the external innovation environment.

'Regional Patterns of Innovation: The analysis of CIS results and lessons from other innovation surveys' is a remarkable survey which shows the diversity of factors and relationships which promote innovation within different regional systems of innovation in the EU. It is based on data from the second Community Innovation Survey (D'Agostino 2000; Step Economics 2000) and outlines three types of successful regional systems of innovation driven by company cooperation networks, state intervention, and academia:

- The first type covers Austria, most of Germany (except Brandenburg and Sachsen-Anhalt in eastern Germany, Oberpfalz in Bayern, Koblenz, Köln, Hamburg, and Braunschweig) and Ireland. This type is characterised by a high diffusion of innovation activities: the share of product and process innovators is very high, as is the number of patent applications. Innovation is targeting cost reduction mainly. Enterprise R&D efforts are at European average level and public financing for innovation is slightly lower. Companies use all available information resources about innovation, whether public or private, and collaboration with universities and research centres is extremely frequent. This type 'looks like a model of innovation diffusion, although the data are not good enough to allow for an underlying coherent innovation system – if any – to be fully described. Anyway, it points to the importance of co-operation between private firms and research institutions, in which firms are not simply passive beneficiaries, but pro-active players who undertake most of the R&D themselves' (Step Economics 2000, p. 10).
- The second type covers two German regions (Hamburg and Koblenz) plus eight UK regions along the length of the north–south axis. It is characterised by high quality human resources, low youth unemployment, small export, small R&D in-house teams, and higher than EU average innovation diffusion. Enterprises develop innovation in order to respond to standards and regulations, as well as to cut production costs. Communication between enterprises is good, but partnership is extremely limited. The lack of financing capital is of primary importance, which appears to be a primarily British problem; this is a paradox given Britain's and London's strengths as a financial centre. State intervention seems to lead to innovation. This type corresponds also to a successful model of innovation diffusion, but this innovation system looks more *government-pushed*, while the previous one looks more *cooperation-driven*.
- The third type covers the greatest part of Sweden and can be characterised as the 'Swedish model: academia-driven innovation'. Research and development expenditure by not-for-profit research institutions (often government-run) is impressive, and significantly higher than EU average. Nonetheless, this knowledge is not always transferred to enterprises, and public research institutes do not collaborate with the private sector, a fact which underscores the weaknesses of the technology transfer mechanisms here. Business expenditure on R&D is also significant, three times greater than the European average. 'The Swedish model is thus a model of high public expenditure (via research institutes), that are not transferred efficiently to the innovative performance of the firms. It seems that the main culprit has to be found in the transfer mechanism that appears particularly poor' (Step Economics 2000, p. 10).

4 **Going global.** Digital space, telecommunications and the Internet contribute to extending the regional systems of innovation to a global scale. Innovative clusters and agglomerations, institutional mechanisms and digital

spaces are intertwined and generate a multi-level innovation system which we have labeled as 'intelligent city/region' (Komninos 2002). Intelligent cities are systems of innovation combining innovative clusters, technology learning institutions, and digital innovation spaces. In this case, the innovation system is constructed on three overlapping spaces: physical, institutional, and digital.

Digital spaces and online knowledge management tools open up new possibilities for product and process innovation. R&D is sustained by global academic and research networks; the supply chains are better coordinated by digital applications of supply chain management linking inputs from all over the world; product development partners scattered in every corner of the globe may interact in real time; the voice of the customer is better heard; end-users of innovative products and services can participate in product design, development, and testing. Digital interaction sustains the core characteristics of innovation systems, such as communication, interaction, cooperation, and joint initiatives.

One of the regions of innovation excellence that has significantly invested in this direction is Uusimaa. It covers the southern section of Finland, and includes Helsinki and 22 smaller towns. The tertiary sector dominates in the region, with more than three-quarters of the population employed in it. During the 1990s, productive restructuring in the region was based on the extensive introduction of information technology and innovation development, and since then these factors have constituted the regional engine of growth. The region has a very strong R&D infrastructure. Uusimaa attracts around 40 per cent of TEKES resources, the major Finnish technology organisation. There is also a significant concentration of universities, with the University of Helsinki being the largest research institute. Others include the Technological Research Centre of Finland (VTT), and the Pulp and Paper Institute, many Polytechnic schools, as well as a large number of incubators (20) (RITTS Helsinki). Today R&D expenditure amounts to more than 3 per cent of GDP and the most rapidly developing sector is telecommunications. Multiple public digital spaces offer services and information and programmes such as Trident, Infocities, Virtual Helsinki, and Helsinki Arena 2000. The public administration has set as main target making the region the leading European city in terms of digital culture and services, and in terms of the degree of useful technological services offered using digital media.

The emergence of global innovation networks that crisscross established innovation patterns was the subject of a recent report by the Economist Intelligence Unit (2007), highlighting new approaches to managing global research and development projects. The report points out that the traditional innovation process of centralised company funding and R&D is gradually being superseded by networks structures seeking ways to disaggregate R&D departments and distribute the innovation process across a network of external partners and offshore sites. The report is based on a survey of over 300 executives worldwide, as well as a series of in-depth interviews with executives and innovation experts, and identified some key attitudes and trends: (1) innovation is becoming more expensive and a solution is to disaggregate the R&D function worldwide and share the burden of innovation with external organisations; (2) R&D is increasingly being moved offshore, with India offering the best combination of cost and quality; (3) innovation becomes increasingly 'open', a trend which reduces time-to-market and maximises value for every organisation involved; and (4) disaggregating R&D worldwide creates major management challenges on intellectual property and communication. Online technologies can aid this process, but face-to-face interaction will remain vital to strategic alignment.

Technological intelligence, market and technology monitoring, online communities of practice, digital knowledge networks, online information and technology collaboration services are some of the applications which may improve global partnerships, systemic networks and the ability to obtain information from sources all over the world.

Discussions about the regions of innovation excellence in the EU and exploring the structural aspects of excellence brings to the surface certain very interesting conclusions about the new regional development model of the European regions.

In order for a region to achieve high performance levels in all innovation indicators it is not enough to attract enterprises with high innovation performance. In parallel, a rich regional environment in terms of population education, innovation institutions, and dissemination of the information society is needed. Excellent enterprises are not sufficient to ensure excellent regions. Rather, the reverse is true: excellent enterprises flourish in excellent regions.

The presence of a powerful innovation environment around the companies is a critical variable for regional innovation excellence. The meaning of this term (innovation environment) cannot be reduced to the conventional concept of 'external economies'. For the majority of enterprises, and particularly for smaller enterprises that do not have their own R&D departments, this environment is the sole guarantee of their innovation capacity, particularly in the case of new product development.

The foundations for generating such environments are to be found into the knowledge-intensive clusters of the region. This is valid to a greater or lesser extent in all regions of excellence. But clusters are complemented by institutions supporting research and development, financing, and technology transfer, as well as by service providers for consulting, corporate planning, cooperation management, product marketing, and other high-value services. Internet dissemination, the use of computers and the presence of digital cooperation spaces also contribute significantly and present a strong statistical correlation to innovation and competitiveness. Complexity is great and the paths leading to innovation are many. Different enterprises in the region can operate within different innovation environments.

The most interesting conclusion, however, relates to the speed of change. In less than 40 years in Germany and 10 years in Finland we have discerned a radical change of traditional agricultural regions, which thanks to the development of strong local innovation systems and ICTs now figure at the top of the European regional ranking in terms of innovation excellence.

3 Systems of innovation
Continuous spatial enlargement

Widening systems of innovation

It is a commonplace that the world is rapidly changing. A new economy is emerging from the opening up of national borders, globalisation, and the application of information and communication technologies in manufacturing and services. Whatever term we use to characterise this new regime of accumulation and development, 'new economy', 'knowledge-based economy', 'innovation economy', it is certain that it stands on the association between R&D, innovation, and competitiveness, which means that the creation and application of knowledge are among the most important drivers of contemporary development and wealth.

The new configuration of development is based on innovation, research, and technological capabilities. However, the spatial distribution of these new factors of development (knowledge, R&D, innovation, high-tech activities, patents, etc.) is much more uneven than the geographical distribution of GDP and development. Regional technology and innovation gaps are sharper and more profound than development gaps. This is the major contradiction of our time. While knowledge-intensive activities, high technology industries and technological innovation are central forces of development, the geographical agglomeration of the new innovation-driven economy is extremely uneven. The drivers of contemporary development and wealth are located unevenly in a few territories and localities. Market forces and agglomeration economies tend to cluster technological innovations into a few islands.

All regions are trying to cope with this challenge, seeking to improve their position in the innovation economy, advance their innovation performance, and increase their share of innovation and high-tech activities. However, there is no universal formula on how to achieve these goals. It is crystal clear that it is not sufficient for a region just to replicate the path that another region has followed. Even the most successful regions go through successive waves of growth, decline and restructuring shaped by continuously changing products, technologies, and innovations. There is no universal formula on how to become and remain innovative.

In this chapter we will try to explain how regions sustain innovation; how different types of innovation systems are formed at regional level; how these systems work as collective mechanisms of learning and innovation; and what principles determine their governance. The relationship is twofold: innovation contributing to regional development, and regions contributing to innovation development.

We will focus on territorial systems of innovation: their diversity and evolution shaped at three successive stages. Small innovation systems, based on physical proximity within clusters, have evolved into larger-scale regional systems due to institutional agreements and regional policies; then with the introduction of advanced information and communication technologies they have become more intelligent and further enlarged to wider supra-national and global scales. It seems that a continuous spatial enlargement of innovation systems is taking place. What is of interest here are the elements constituting each successive stage, the internal mechanisms of innovation production, the emerging character of the respective territorial system of innovation, endogenous or exogenous formation, the barriers to development and characteristic cases as well.

We will also look at the governance characterising each stage: local alliances with a strong presence of chambers of commerce and industry and other sectoral agencies in the case of clusters; regional authorities, national programming authorities, and the European Commission in the case of regional systems of innovation; multiple associations and private–public partnerships in the case of intelligent global systems of innovation.

It should be stressed that the policy paradigm deriving from the theory of territorial systems of innovation is proposed equally for advanced and less favoured regions. However, when it comes to the question of transfer of practices and policy models from technologically-advanced to less-developed regions, the main challenge is whether this happens in an uncritical and non-informed manner or whether it is targeted at the creation of endogenous learning and innovation mechanisms taking into account local specificities, strengths and weaknesses.

'Each year the US spends more than $260 billion on "innovation" (as measured by R&D expenditures) – globally the amount spent on R&D exceeds $600 billion per year. The outcome of this investment in innovation is prodigious. Over 100 new patents are applied for each hour, and 2,265 new businesses are started each day. An estimated 80–100,000 new products are introduced in the US alone each year. But the results are dismal at best. Each day 9,180 businesses fail, a 96.25 per cent failure rate. Depending on the source, anywhere from 80–95 per cent of new products either fail outright, or fail to meet their business objectives.'

Source: www.alwayson-network.com

Innovation

Let us start from the concept of reference. The term 'innovation' denotes the act of starting something for the first time; introducing something new; a creation of a new device or process resulting from study and experimentation; the creation of something in the mind (WorldNet dictionary). In Webster's dictionary, it is defined as the act of innovating; introduction of something new, in customs or rites; a change affected by innovating; something new and contrary to established customs, manners, or rites.

However, an important aspect is missing from these definitions. Innovation is not only new, but novelty combined with better performance or efficiency. 'New' in itself is futile unless it gives an additional advantage, a better solution to a problem, a cost reduction to an operation. Whatever you consider, be it computers, drugs or transport, innovation and new products always bring some additional use value: more rapid chips, more efficient or less harmful drugs, more environmentally-friendly and safer cars, transport at lower cost.

The concern about performance becomes clear in the OECD definition of innovation:

> A technological product innovation is the implementation/commercialisation of a product with improved performance characteristics such as to deliver objectively new or improved services to the consumer. A technological process innovation is the implementation/adoption of new or significantly improved production or delivery methods. It may involve changes in equipment, human resources, working methods or a combination of these.
>
> (Oslo Manual 1995)

Typology is a good starting point for understanding what innovation is. Typology reflects an empirical wisdom and a classification of events. From a typological point of view a classical distinction is drawn between product, process, and organisational innovations:

- *Product innovation* is linked to the development of new products and the rise of new industry sectors; in these early stages many small companies co-exist into the industry and there is a lot of product experimentation. As industries mature, standardised product designs appear and dominant production and marketing models are formed.
- *Process innovations* relate to the use of more advanced production technologies and become more important as volume rises and industries search for economies of scale and scope. Further increases in the company size turn interest outside the company, upstream and downstream of the production site.
- *Organisational innovations* (often referred to as non-technological innovations) introduce more efficient cost arrangements by re-organising the entire

supply chain through lean production, co-operation networks, flexibility, just-in-time delivery systems, and optimisation of producer–supplier relations.

Any change in product and processes is not necessarily innovation; equally all innovations are not the same. Small changes in product form and appearance are not considered innovations. A change in the knowledge base of products or processes is necessary to certify an innovation. From the point of view of importance, innovations can be roughly classified into three main types: *incremental innovations* which characterise small improvements along the learning curve and 'learning by doing' – these improvements are continuous and future changes can be predicted with confidence; *radical innovations* where a totally new technology comes along and displaces the incumbent technology, as happened in the case of the optical recorder replacing the VCR or the transistor replacing the vacuum tube – these changes are discontinuous and cause 'creative destruction' in the industry concerned; and *general purpose innovations*, which mark the entire technological regime, such as steam power, electricity, internal combustion engine, computers, and the Internet, which affect a wide range of industries, products and processes.

Innovation is radical when it relates to the development of a new product or solves a problem for the first time. On the other hand, innovation is incremental when it relates the solution to a problem already provided by another organisation or in another area. Incremental innovation is also important, despite the fact that it solves a well-known problem. The fact that a problem has been solved in the past by some other organisation or in another geographical region is no guarantee for the success of the solution when applied under different conditions to different subjects. This discussion, down to the principles to deductive reasoning, is widely known in the philosophy of science by Bertrand Russell's tale of the deductionist turkey.[2] Innovation was always associated with the unpredictable, with overturning established trends, with maneuvers at the limit of and outside the rules.

Christiansen introduced the term 'disruptive technologies' to characterise a low-performing, less-expensive technology that replaces an established technology; the issue was discussed at the Disruptive Tech NACFAM Workshop, at which several disruptive technologies were identified: advanced sensors, micro-fabrication, modelling and simulation, reconfigurable tools and systems, smart systems, solid free-form fabrication, visualisation and planning, and knowledge management (Malone 2005).

Two important modern aspects of innovation are collaborative innovation and continuous innovation. The first refers to innovation networks and collaboration in developing new products. It is identical to the concept of 'open innovation', which describes a network of innovation organisations that collaborate in new product development. The second refers to the concept of new product platform, a core platform, from which many product prototypes derive either at the same time or over time. The core platform is adapted to various markets and users requirements. In this case, innovation is continuous and products are developed over time adapting to market changes and technology updates. It is extremely difficult to copy a stream of continuous innovations, since there is not one product to copy but a flow of new, ever changing products.

Towards systems of innovation

The work of Josef Schumpeter has influenced deeply the understanding of innovation. He distinguished five types of innovation: (1) introduction of new good or a new quality of a good; (2) introduction of a new method of production; (3) carrying out a new organisation in the industry; (4) opening up of a new market; and (5) the conquest of a new source of supply of new materials or semi-finished manufacturing parts. However, his work was much more than classification.

Schumpeter's original theory of innovation emphasised the role of entrepreneurship and small companies in seeking out opportunities for novel value-generating activities and profit. He introduced the distinction between 'invention-discovery' and 'innovation-commercialisation'. The separation of invention from innovation characterised the typical innovation model of the late nineteenth century, in which independent inventors provided new product and process inputs to entrepreneurial firms. Later, Schumpeter became aware of the rise of in-house R&D departments in large companies, which changed the innovation landscape radically. From this division between early to late Schumpeter comes the distinction between his 'Mark I' model of innovation and 'Mark II' model in which innovation is envisaged as a more routinised process within large companies. The Mark I model is associated with Schumpeter (1934), while the Mark II model with Schumpeter (1943). However, the shift and emphasis on the role of large firms and R&D labs as key agents of innovation may be seen as reinforcing his earlier theory of innovative profits, since large firms may exercise market power. The standard interpretation of the relationship between profits and innovation focuses on the quasi-monopoly conditions created by innovations, enabling innovators to establish a temporary monopoly within the industry, capable of generating super profits because of higher output prices and lower input prices and costs (Cantwell 1989).

In the 1980s the reliability of this interpretation on the birth of innovation was challenged on many sides since it had become clear that innovation is affected by many factors outside a business, regardless of whether it is large or small. The linear model of innovation production directly from research began to be abandoned.

Jaffe (1986 and 1989) found that the innovative performance of firms depends not only on their own investments in R&D, they are also strongly affected by the R&D spending of other firms and universities. But if the ability to innovate is affected by external sources of knowledge and technology then we should expect wide differences in the innovative performance of firms located in different regions.

Griliches (1979 and 1984) developed an input-output model linking patented innovations (output) with new technological knowledge generated by R&D in industries and universities (input). The model has the form:

$$\ln PATs = \beta_1 \ln IR\&Ds + \beta_2 \ln UR\&Ds + \beta_3 \ln Cs + POPs + \varepsilon s$$

where

- lnPATs is the natural logarithm of the number of patents granted to private manufacturing firms in the state S
- $\beta_1 \ln IR\&Ds$ is the natural logarithm of R&D expenditures by manufacturing firms in the state S
- $\beta_2 \ln UR\&Ds$ is the natural logarithm of R&D expenditures by universities in the state S
- $\beta_3 \ln Cs$ is the geographical coincidence index
- POPs is the total resident population in the state S
- εs is a stochastic error
- $Cs = \Sigma ic\ [UNIVic \star Tpic]/[\Sigma ic UNIV^2 ic]^{1/2} \star [\Sigma ic TP^2 ic]^{1/2}$
- UNIVic is R&D expenditure within universities by industry i and metropolitan area c. Tpic is the number of researchers in the manufacturing sector by industry i and metropolitan area c.

Though the model clearly belongs to the linear conception of innovation, attributing innovation mainly to R&D, it also shows that innovation performance (measured by patents) is affected by factors external to the company such as university R&D, the size of population, and the geographical agglomeration of universities and industries. The variables of population and geographical proximity were later strongly re-introduced by the systemic theories, in terms of clusters, skills, and market conditions.

Estimating this model for 29 US states over the period 1972–1977 and 1979–1981 Jaffe (1986) showed that corporate patenting is significantly affected by spillovers from both the private corporate R&D expenditures and research expenditures by universities, although the former (showing elasticity >0.7) have a stronger impact than the latter (elasticity <0.1); the geographic coincidence index is only marginally statistically significant.

Piergiovanni and Santarelli (2001) estimated the same model in the French regions using data for 1991; they showed that spillovers from university R&D are a relatively more important source of innovation in private and state-owned industrial firms than industrial research itself. This is due to the national system of innovation in France and the familiarity of French firms with technological dissemination projects from universities and public research centres.

The impact of external factors on innovation was expressed more plainly with reference to the environment in which companies operate. The evolutionary metaphor formulated by Nelson and Winters (1982) introduced a robust relationship between the internal and external environment of the company:

- Companies follow *organisational routines* which are behavioural patterns inside the firm and ways of doing things in production, R&D, marketing, management, etc.
- Innovation starts by *search activities*, which are organisational activities associated with the evaluation of current practices (routines), searches for more efficient practices outside the company, leading to modification and/or replacement of routines.
- The modification of routines is influenced by an external *selection environment*, which is formed by organisations that affect the transformation of knowledge to products (consulting, marketing, finance, engineering competence).

These fundamental processes of innovation (routines, search, and selection environment) create a cognitive space, which is specific and exclusive to each organisation. Central to Nelson and Winters' (1982) thinking is that technologies set boundaries to innovation patterns; learning processes are dependant on their technological environment, which they characterise as a 'technological regime'. The concept of technological regime more accurately describes the technological environment in which a company operates. They identified two technological regimes: an 'entrepreneurial regime', associated with scientific research, where new innovative firms can easily enter; and a 'routinised regime' which characterises innovation of established firms, having a cumulative knowledge base. Further research on this issue defined the main components of technological regimes: (1) opportunity conditions, which reflect the probability of innovation at any given amount of resources; (2) appropriability conditions, which reflect the capabilities to protect innovation from innovation and reaping of profits from innovative activities; (3) cumulativeness of innovation, which denotes the continuity of a technological environment and the conditioning of actual by past innovations; and (4) the nature of knowledge, reflecting the proprieties of knowledge upon which innovation is based (Breschi 2000).

Along the same line Clausen (2004) studying the Norwegian industry distinguished five types of technology regimes:

- The Science-Based regime, characterised by high levels of technological opportunity, technological richness, high technological entry barriers, and cumulativeness of innovation, and directly associated with advances made in academic research. Typical industries in this regime are pharmaceuticals and electronics.
- The Fundamental-Process regime, which displays medium levels of technological opportunity, high technological entry barriers due to economies of scale, and process innovations. Typical industries in this regime are chemical and petroleum industries.

- The Complex-Systems regime, which is associated with medium to high levels of technological opportunity, entry barriers in knowledge and scale, dependence from external sources of knowledge, and a knowledge base combining the mechanical, electrical/electronic, and transportation technologies. Typical industries are the aerospace and motor vehicle industries.
- The Product-Engineering regime, which relies on mechanical engineering technologies. It is characterised by medium to high levels of technological opportunity, low entry barriers to innovation, and not very high persistence of innovation.
- The Continuous-Process regime, which relies on a knowledge base combining mechanical and electrical technologies. It is characterised by low technological opportunities, entry barriers and innovation persistence. It includes a variety of production activities such as metallurgical industries, metals and building materials, chemical process industries, textiles, paper, food and tobacco.

A further push to the turn towards the external environment of innovation was given in the late 1980s and early 1990s with a series of publications on 'National Innovation Systems'. The term was introduced by Freeman (1987) and a few years later Nelson and Rosenberg published *'National Innovation Systems'*, which contained studies of 15 countries all over the world: large market-oriented counties (the US, Japan, Germany, Britain, France, Italy), smaller high-income countries (Denmark, Sweden, Canada, Australia), and newly industrialised states (Korea, Taiwan, Brazil, Argentina, and Israel). The book described the operation of national innovation systems arguing that 'the technological capabilities of a nation's firms are a key source of their competitive prowess, with the belief that these capabilities are in a sense national, and can be built by national action' (Nelson and Rosenberg 1993). National innovation systems continue and advance evolutionary thinking. Systems more efficiently describe the external selection environment influencing the processes of change of organisational routines within the company. In particular, systemic approaches focus on the interplay between institutions involved in the creation, diffusion, and application of knowledge, and lead to a better appreciation of the importance of the framework conditions of innovation, like regulations and policies within which markets operate, and the wider governance of innovation.

All definitions of national innovation systems put an emphasis on institutions: 'the network of institutions in the public and private sectors whose activities and interactions initiate, import, modify and diffuse new technologies' (Freeman 1987); 'the national institutions . . . that determine the rate and direction of technological learning' (Patel and Pavitt 1994); the 'set of distinct institutions which jointly and individually contributes to the development and diffusion of new technologies' (Metcalfe 1995); 'institutions and economic structures affecting the rate and direction of technological change in society' (Edquist and Lundvall 1993).

Narrowly, the concept refers to the network of institutions in the public and private sector whose activities and interactions influence the entire innovation environment: framework conditions (legal, economic, financial, and educational) setting the rules and range of opportunities for innovation; the science and engineering base; the transfer and absorption of technology; and the innovation dynamo, which covers dynamic factors within or immediately external to the firm. On a broader definition, the system includes all parts of the economic system and institutional set up affecting searching, learning, and producing knowledge.

The characteristics of a national innovation system can be summarised in a number of fundamentals: Firms are part of a network of public and private sector institutions whose activities and interactions initiate, import, modify and diffuse new technologies; institutions are the cornerstone of each system; the system is based on linkages (both formal and informal) between institutions; there are flows of intellectual resources between institutions and learning is a key resource; geography and location still matter (Holbrook 1997, p. 5). The emphasis is clearly on institutions, networks, and knowledge flows rather than on the national character and boundaries of these interactions.

From this point of view, other types of innovation systems may also be defined, namely regional systems covering sub-national scales and sectoral systems corresponding to different manufacturing and service sectors. Regional systems of innovation are built upon the local knowledge base, experience, and trust relationships created along the development path of the region in question (Heidenreich 2004). They stem from the accumulation of regional competences and are bound up with the particular history of each region.

Innovation systems explain innovation performance with respect to networks and interactions among companies, universities, and government. These interactions are not the same in all industry sectors. Universities play an important role in pharmaceuticals and computers, but a modest one in aircraft and steel. Government funding is important to some industries and unimportant to others. However, the main question of whether these networks, interactions, and communities have a national character, remains unanswered. Rather, the national prefix to innovation systems has been attributed to the strength of the nation-state intervention, laws, and policy in recent European and US history.

Empirical surveys in CIS-1 and CIS-2 have confirmed a set of obstacles preventing companies from transforming their internal routines and innovating. They originate, by order of significance, from internal factors (lack of qualified personnel, management rigidities, information on markets), financial factors (high cost of innovation, limited funding sources, high risks), and market factors (limited customer response, standards and regulations). It is within the innovation system that companies will search for resources to fulfill these needs in terms of knowledge, funds, and market research. The system counterbalances individual weaknesses.

What makes an innovation system?

The concept of innovation system refers to a 'perpetual cycle borne out of the critical inputs of intellectual and financial capital, translated into new technologies and products that lead to new firm formation and job creation, generating revenues that may be re-invested into the system' (John Adams Innovation Institute 2007). In order to make clear this internal dynamic of an innovation system it is necessary to define: (1) the elements of the system, the subjects, organisations, and institutional bodies which participate in it, and shape the initial conditions for system start-up and operation; (2) the relationships between the elements, which define system operations, internal processes and transformations, inputs and outputs; and (3) the regulatory mechanisms and governance requirements that ensure that the system operates and reproduces itself.

Elements

In innovation systems theory the key building blocks of the system are institutional entities: firms, research organisations, funding, and technology intermediary organisations. Today it is fully accepted that innovation systems can be local, regional, national, international or global depending on the geographical spread of the elements comprising each system at national and international level. Since the number of elements comprising an innovation system is usually very large, they are expressed via categories such as firms, suppliers, research labs, financing organisations, technology transfer organisations, etc. The system is conceived of categories rather than individual organisations. In cases where the innovation system is referred to in quantitative terms, the elements comprising it are usually also referred to by quantitative indicators that define the initial start-up conditions or system status at a specific time juncture.

Relationships

Relationships between system building blocks determine how it operates. For example, in Griliches' model, relationships between industrial research, university research, the population, and the geographical proximity of industry to universities determine the innovation performance of an area that is measured in the number of patents filed for. For there to be a system, relationships between its constituent elements should be stable and long-lasting.

Relationships between constituent elements express the transformations that occur within the system and the innovation generating mechanism that emerges from the synergies between its elements. For example, a web of relationships interconnects invention, innovation and commercialisation practices. Invention is the creation of a new idea or concept; innovation is taking this idea to make a product or process, and change it into commercial success. The stages of this entire process include invention, translation to innovation, and commercialisation. Most successful innovations are based on inventions coming from R&D

and science and technology. However, in many cases initial ideas and stimuli also come from fashion, art or the business world. The important issue is the capacity to transform an idea into a product and make it a commercial success.

The cohesive substance in innovation systems, the substance that connects the elements of the system, is knowledge. Organisations constituting an innovation system exchange knowledge and alter the organisational routines for production and exchange. New products and services emerge from knowledge transformations. Innovation is a process of ongoing improvement and enhancement of knowledge, which includes procedures to record knowledge, to transfer knowledge from other scientific or technological disciplines, to assimilate existing knowledge, to establish knowledge complementarities and recombine old and new knowledge, to develop knowledge application capability, and disseminate new knowledge embedded into products. This process, which is shown in Figure 3.1 is evolutionary in the sense of a continuous development of know-how, but it is not linear, as shown in this diagram.

The process of changing routines, which is described by the evolutionary paradigm, is represented in this figure as an evolution of existing knowledge status via knowledge transfer, development, assimilation, and dissemination. A starting point for transformation is to formulate a question or problem that needs to be solved. It may relate to a more efficient production system, to a more durable engine, a pharmaceutical with fewer side effects, a new energy source, a material or labour saving device. The first step in the process is observation, monitoring, and assessment. Observation and understanding are not the same thing. Secondly, learning requires an apprenticeship relationship in which questions brought to light by observation are interpreted. The first answers are given by the assimilation of existing knowledge. Thirdly, new research is needed to deal with aspects of the problem for which satisfactory answers have not been provided. Fourthly, theoretical knowledge should be transformed into applied knowledge embodied into objects. Fifthly, along with

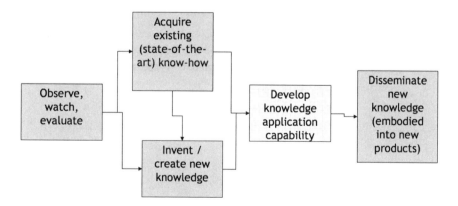

Figure 3.1 Knowledge evolution along the innovation process

presentation of the final product or new method, the new knowledge it embodies is disseminated. Knowledge assets corresponding to individual stages in this transformation may belong to different subjects and organisations. Within innovation systems, such knowledge fragments are integrated.

Governance

The connections between constituent elements in an innovation system are interspersed with institutional switches which regulate the flow of knowledge between system elements. These are intermediation, technology transfer, and financing organisations which make it easier for interfaces to form or prevent them forming. This form of governance is direct and active. However, governance cannot under any circumstances be limited to direct and active intervention only.

At the same time, a set of institutions, standards, formal and informal rules and norms, regulate how the innovation system operates and ensure its emerging behaviour. Proper operation of an innovation system can be attributed to the spontaneous, unseen operations of bottom-up regulatory mechanisms. This is the meaning of Nelson and Rosenberg's remark on innovation systems:

> Although to some the word (system of innovation) connotes something that is consciously designed and built, this is far from the orientation here. Rather, the concept is of a set of institutions whose interactions determine the innovative performance of national firms. There is no presumption that the system was, in some sense, consciously designed, or even that the set of institutions involved works together smoothly and coherently. Rather, the 'systems' concept is that of a set of institutional actors that, together, play the major role in influencing innovative performance.
> (Nelson and Rosenberg 1993, p. 3)

Frequent reference to the important role of culture and emerging trust in clusters refers to the importance of indirect regulation of collaborative relationships and networks.

Enlargement I: cluster-type innovation systems

The first analytical account indicating that innovation emerges from a system of companies rather than the individual company appeared in late 1970s writings about industrial districts. In 1977 Bagnasco published his study on the Third Italy describing 'system-areas', communities and cities of central Italy flourishing on the basis of local vertical integration among small companies belonging to the same industry. Michael Porter popularised the concept of industry clusters in *The Competitive Advantage of Nations* (1990), recognising that at regional level the majority of economic activity takes place in a limited number of industry clusters/sectors.

The meaning of a cluster is the spatial concentration of industries that gain advantages through location in proximity and agglomeration economies, either of scale or scope. Interconnected companies and institutions develop systematic relationships to one another based on complementarities or similarities in particular fields, cooperate and establish close linkages and working alliances to improve their competitiveness. Over the last 25 years, since the opening up of this discussion, clusters and industrial districts have continued to occupy a central place in the debate about regional innovation.

Clusters have different origins: many (Italian districts) have grown based on the voluntary decision of manufacturing SMEs, while others have been influenced by large manufacturing companies (Bayer in the Rhine region), and others are by-products of universities and research institutes, in the case of science and technology parks. The classical typology of clusters is given by Porter, who draws the distinction between vertical vs. horizontal clusters:

- *Vertical clusters* are clusters with strong inter-firm linkages along the supply chain; the companies are specialised in different phases of the production process, and linked with supplier–producer relationships; characteristic cases are the Italian industrial districts.
- *Horizontal clusters* are clusters with weak inter-linkages; the organisations comprising the cluster act as a whole (as agglomeration/swarm of organisations) to achieve a common objective, i.e. to open a new market, to use infrastructure, to cooperate with a strong R&D institution. Horizontal clusters also include industries, which might share a common market for the end product, use a common technology or labour force skills, or require similar natural resources.

Other typologies have also appeared based around the content of clusters:

- *Industrial districts* in traditional sectors, such as food, paper, plastics, mechanics, jewelry, leather, shoes, furniture, clothing.
- *Technologically-advanced districts* in high-tech sectors of electronics, computer related, telecommunications, biotechnology, shaped by applied research, venture capital, and specialised services.
- *Planned clusters*, science and technology parks, focusing on technology transfer and/or the attraction of high technology activities and investments.

Or with respect to a life-cycle point of view:

- *Potential clusters*, just an agglomeration of organisations, characterised by the concentration in a region of a number of firms and other actors, but with low density business networks and lack of synergies.
- *Emerging or embryonic clusters* characterised by a certain concentration of firms at local level, which are starting to cooperate around a core activity and

to realise common opportunities through their linkage, but still have low awareness of the potential gains from clustering and low synergy between them.
- *Established or mature clusters,* that is to say clusters that have reached a critical mass of actors highly involved in a region, developed relations inside and outside of the cluster, an internal dynamic of new firms creation and use of common infrastructures and services.
- *Declining clusters,* clusters that have reached their peak and have to adapt to changes in order to be sustainable.

The actors in cluster initiatives belong to many different categories:

- Firms (SMEs, large firms, business champions and leaders, business start-ups and spin-offs), which are the central clusters' actors. In many cases large firms can play a leading role, especially in the initial phase of the cluster.
- Research and educational organisations (R&D institutes, universities, public research labs), which have a supporting role in the clustering process, driving innovation and network creation, and setting main nodes of cluster development.
- Financial institutions (business angels, banks, venture capital, trading houses, investor networks, and other business service organisations) that provide seed capital, financial advice, support the inner dynamics of clusters, management support to start-ups and spin-offs.
- Institutions for technology intermediation focusing on the exploration of methods, diffusion of entrepreneurial culture, cooperation and trust building. These are formal or informal (network) organisations which promote the cluster initiatives among the actors involved and favour the diffusion of the culture for cooperation and trust building.
- Policy-makers (governments, local authorities, regional development agencies). They play a catalytic role, having broad visions and goals, and providing legitimacy, support mechanisms and infrastructures.

As a system of innovation, the cluster is characterised by the way the aforementioned actors are connected and cooperate. The first account on the innovation mechanism of clusters and the way that clusters promote innovation capabilities of their members was provided by Becattini (1989). He described the creativity of industrial districts in terms of **specialisation and agglomeration of skills**: Within the district/cluster concentrate many and diverse skills covering various fields of knowledge and production. Even in cases where the whole cluster focuses on a single industrial sector, the multiplicity of skills comes from specialisation of the cluster actors in different stages of the production process. Cooperation networks are created among the members of the cluster and the skills they have. Innovation stems from the combination of skills, knowledge, and resources that are put together. However, to this end, a minimum of cluster members is necessary; 100 organisations, for instance, have been considered as a threshold for defining a production complex as an industrial district. Within the district catalysts of interlinking transactions are facilitating networking among the diverse skills and production units. In the case of industrial districts in Tuscany, this role is played by the 'impannatori', who constantly re-organises the supply chains of the district in relation to the orders they get. (They may also provide a model for the incorporation of small firms in larger global supply chains.) Venture capital or business angels also act as a catalyst in high-tech clusters, as do the technology transfer centres and the liaison offices in the case of technology parks. Along the same line, Paci and Usai (2000) have argued that as firms gather together in industrial districts it is likely that the locality gains useful infrastructures and an appropriate specialisation pattern facilitating the provision of goods, production factors and services.

Agglomeration is also important for **knowledge spillovers**. Knowledge, which is the prime base of technological change, is highly volatile and can be easily appropriated by other firms in a specific area. Jaffe et al. (1993) also made a clear argument in favour of knowledge spillovers within the clusters, arguing that spillovers, as measured by patent citations, are mostly likely to occur within geographically bounded areas rather than freely flowing across regions.

A different explanation of the innovation mechanism of clusters came from Lawson and Lorenz (1999) who attributed it to **collective learning**: the capacity of firms within the cluster to acquire and generate new knowledge. The concept of collective learning had been initially advanced by GREMI to connote the capacity of a particular regional innovative milieu to facilitate innovation by the firms that are members of that milieu and to reduce the uncertainty created by rapid technological change (Keeble et al. 1999). The concept describes the phenomenon that regional clusters of SMEs develop a capacity for self-sustaining technological learning, innovation, and new product development. For Camagni (1991), who spoke explicitly about collective learning, the concept focuses on links and networking between firms via the local labour market. A survey in the area of Cambridge (UK) has attempted to explain the innovation capability of this region with respect to the above ideas, and identified three regional collective learning processes: (1) spin-offs and start-ups by Cambridge University and large R&D consulting companies; (2) inter-firm cooperation and networking with suppliers, subcontractors, service providers, research collaborators on local, national and international scale; and (3) skilled labour mobility within the local labour market, especially of scientists, engineers, research staff and managers (Keeble et al. 1999).

A systemic explanation of the innovation mechanism of Italian industrial districts is also given by Poti and Basile (2000). Innovation system analysis seems to be relevant in determining the probability that firms introduce new products on the market and they help in discriminating among low and high developed regions. The authors developed a quantitative model to explain divergences in region/country propensity to innovation through a system of innovation approach. Starting from the idea that externalities and spillovers have positive effects on the innovative performance of firms, they focused on three types of interactions: (1) collaborative and network inter-firms relations; (2) sectoral-regional clusters; and (3) inter-institutional (industry-public research institutes) relations.

The model of innovation generation has the form:

$$INNOVATION_{ij} = + SIZE_{ij} + MARKET\ INCENTIVES_{ij} + TECHNOLOGICAL\ REGIME_{ij} + ORGANISATION_{ij} + SPILLOVERS_{j} + PUBLIC\ R\&D_{j} + PUBLIC\ SUPPORT_{ij}$$

where i indicates the firm and j indicates the region.

Variables on inter-firms organisations, local (subregional) spillover and policy were added to the Schumpeterian equation on innovation determinants. The model tested the significance of organisational variables, externalities and public policy variables in explaining differences in firm innovation propensity among less and more developed regions in Italy. The model showed that the relationship between innovation and firm organisation differs among regions. Local spillover variables have a significant impact on the firm's propensity to innovate at national level, and it also discriminates among regions. Public support for innovation plays a different role in different regions.

Spontaneous clusters

Sustaining the innovation capacity of existing clusters is more feasible than creating new clusters. A cluster is far more than a simple club of companies or a supplier–customer network. Most of the innovative capacity of clusters is emerging, thus making planning and top-down formation extremely hard. Apart from gathering enough companies to have a critical mass, a cluster has to meet collective innovation objectives. However, even in spontaneous clusters four major areas of cluster building practice can be identified.

- *Defining strategy and vision*: Defining cluster strategic direction, defining steps to take, defining assessment methods, setting up formal clustering organisations. It also includes the generation of identity and building a critical mass (attracting new-coming companies, withholding the ones already settled within the cluster and on the territory) at national and international levels.
- *Building social capital and creating trust*: Preparing the ground for collaboration, building and nurturing trust, sustaining trust; resolving collective and collaboration problems; sharing of means in terms of technology, human resources, training, strategic watch, quality follow-up, technology transfer are the critical issues here.
- *Sustaining strategic linkages*: Formalising linkages, obtaining structured routines for interaction; negotiating capacity improvement towards customers, OEM, and suppliers so that they integrate more value adding functions; meeting the demand of OEM to reduce the number of first rank subcontractors.

- *Management of external knowledge flows*: The cluster has to be closely connected to the R&D and technology transfer set ups. It also means undertaking cluster actions which aim to improve cluster dynamics (new technology and firm growth, inter-actor network creation, cluster formation) and cluster environment in terms of markets and technologies available within the cluster.

Planned clusters: Science and Technology Parks

Technology Parks (a term also covering Science Parks, Research Parks, and Innovation Centres) offer the simplest way to plan innovative clusters. According to calculations made by Anttiroike (2004) there are around 500 technology parks in the US, 400 in Europe, 120 in China, 120 in Japan, 40 elsewhere in Asia, and 60 in other regions of the world. It is a very popular technological development institution because it offers the simplest way for achieving innovative cluster or technology districts. The bottom-up, emergent, complex innovation mechanism that operates within a spontaneous cluster or industrial district is held by a top-down mechanism, which brings a relatively small number of research, technological intermediation, and entrepreneurial organisations into contact, creating a technology transfer and knowledge utilisation web. The achievement of science and technology parks is that they permit the planned construction of an innovation cluster, transforming the chaotic dynamic of technological collaboration within clusters, into simpler technology transfer relationships.

The establishment of an innovation mechanism within a Technology Park is based on the synergy between four elements:

- Demarcating the area and developing suitable infrastructure.
- Creating a research core that is usually comprised of research institutes or university research labs. The scientific and technology specialisation of such organisations determines the specialisation of the Park and the technological innovation mechanism that operates in it.
- Creating a productive/business core comprised of small enterprises, spin-off businesses newly established by research institutions, and larger units or R&D departments of multinationals attracted to the Park to make it easier to collaborate with research institutions or use existing research infrastructure.
- Creating a technology transfer core which is comprised of technology transfer organisations, consulting companies, technology centres focused on specific technologies and incubators. All these organisations intermediate in their own different way in transferring results of research to businesses.

Among these elements, collaborative relationships initiate different types of innovation mechanisms such as: (1) joint development of new products between companies and research labs; (2) technology transfer from research organisations to product manufactures and service providers; (3) spin-offs which seek to commercially exploit research results; (4) attracting knowledge and technology intensive organisations; and (5) increases of land value due to the brand name created by concentrating technological organisations in the Park and reinvestment of revenues in technology infrastructure and services.

These aspects determine the innovation governance in planned clusters which extends to four different areas of management:

- *Innovation and technology management*, focusing on R&D exploitation; technology transfer, licensing, and IPR; promotion of technology platforms; and creation of technology transfer centres and networks.
- *Spin-off management*, focusing on the construction of incubators; provision of seed capital/equity capital; provision of innovation services to incubatees; and other forms of spin-off support.
- *Attraction of external organisations*, including the definition of target groups; global marketing; provision of incentives and attraction packages; and aftercare services.
- *Land and infrastructure management*, ranging from acquisition of land; definition of planning regulations; infrastructure creation; to land promotion, and construction.

However, planning and setting in motion a cluster-type innovation mechanism within a Park seems extremely difficult. Several recent studies have concluded that Science Parks tend to fail in attracting and developing high-tech companies and have therefore not fulfilled their expected role as catalysts of regional economic growth. Recently, based on two in-depth case studies of science parks in Denmark and the UK, Hansson *et al.* (2005) discussed alternative mediating roles for science parks in the science–industry relationship. Their suggestion is that the new role of science parks may be to cater for the development of the social capital necessary for enabling and facilitating entrepreneurship in networks.

My interpretation of the frequent failure of Technology Parks as innovation generating mechanisms is exceptionally straightforward. It is based on the logic of innovation systems as emerging mechanisms, which are highly complex, cannot be modelled in full, and initiated via top-down mechanisms such as planning and innovation policy.

Small clusters within incubators

Small innovation clusters emerge within business incubators. An incubator offers business set-up and hosting facilities, infrastructure, services and financing on an interim basis. In addition to providing accommodation, they also offer support services for new product development and technology issues (as well as legal, organisation, and marketing issues) and part of the coaching necessary to 'grow' the business.

The small clusters that emerge within incubators are horizontal. The relationships forming the innovation system are like a constellation, at whose heart lies the incubation mechanism and at the perimeter there are small businesses housed at the incubator. The synergistic relationships between the organisations in the cluster are ensured by venture officers who participate in the management and administration of each incubating business as representatives of the incubator. Usually each venture officer monitors and co-manages three to four businesses as they grow to maturity, until they acquire the organisational and business autonomy to break away from the incubator.

As an innovation system, an incubator gives birth to innovation within a new business development cycle that lasts around 24 months. The lifecycle of a potential innovation which becomes part of this system goes through the following stages: (1) Screening and acceptance; (2) Business planning and seed funding; (3) Company generation; (4) Business mentoring; (5) Product launch; and (6) Graduation and first round of VC funding.

Each stage involves specific activities to be implemented to enable the business to move on to the next one. The cycle starts with nothing more than a promising idea for a product, service or technology and ends with a fully developed organisation that can produce and offer a new product.

The effectiveness of the innovation seeding mechanism in this system is dependent on several factors. Firstly, on the knowledge and technology reserve from which ideas about new products and technologies are drawn. This reserve depends on the geographical area from which ideas are drawn; the larger the area, the larger the technology options available. Secondly, the evaluation mechanism that selects which ideas will be included in the funding and support process. Thirdly, coaching services to support the development of new products/services.

The evaluation criteria – an exceptionally critical element in the innovation mechanism – include assessment of innovation, technological viability, the market, and management capability of the business (Table 3.1).

Table 3.1 Evaluation of innovative start-ups

Evaluation object	Criteria
Initial idea for product(s)	1 Innovation of product/service 2 Similar products/technologies already on the market 3 Copyright, patent, information disclosure
Technical feasibility	4 Technical feasibility 5 Production planning 6 Strategic alliances/cooperation
Target markets	7 Usage value of the product 8 Market analysis 9 Competition strategy 10 Marketing/promotion plan
Start-up person/company	11 Commitment of entrepreneur 12 Maturity of the business plan 13 Capacity to protect innovation 14 Funds available

These criteria ensure that the start-up will produce an innovative product that is technically viable, with an adequate market, and reliable management. Evaluation is carried out by a number of evaluators to limit the subjectivity of the evaluation process. The decisions are independent and of equal worth. There is no provision for consensus among the evaluators. Thresholds are set for certain criteria (such as technical viability or funding) which will lead to a negative decision if not satisfied. After initial selection, the development of innovation is determined by the incubator's coaching ability, its ability to offer business intelligence services, product development services, and marketing support.

Enlargement II: regional systems of innovation

In the 1990s innovation theories turned towards learning organisations and regions, while policies started experimenting with regional innovation strategies. The focus has clearly shifted to learning institutions and regional systems of innovation.

The shift from technology districts and clusters to regional systems of innovation and the 'Learning Region' paradigm was fuelled by three theoretical transitions:

- *From inter-firm cooperation within districts to learning networks:* The contribution of the District theory, wrote Lawson and Lorenz (1999), was more in the area of understanding the territorial foundations of inter-firm cooperation than in understanding the contribution of territorial clustering to a firm's capacity to learn and generate new knowledge.
- *From individual to organisational learning:* Individual learning refers to the acquisition of information, knowledge, understanding and skills, through participation in some form of education and training, whether formal or informal. Organisational learning depends upon individual learning and builds upon it. Organisational learning amplifies the knowledge created by individual organisations, by appropriating knowledge from outside or by creating new knowledge in interaction and collaboration to other organisations.
- *From linear to systemic theories of innovation:* A process (innovation) hermetically sealed within the company and the industrial research lab started to be viewed as system that covers an entire city-region connecting actors from the finance, technological, and production communities and academia.

The region was conceptualised as a living organism with technology learning, management, selection, and knowledge development capabilities. Innovation is based on a system of clusters and institutions in the fields of R&D, tech transfer, finance, technological information, and production. The system contains both demand and supply institutions, and integration is due to knowledge, financing, and marketing networks. Networks within the regional system allocate 'formal' and 'tacit' knowledge and enable collaboration and joint efforts at three different levels: in the interior of clusters, between clusters and innovation support institutions, and between R&D and technology intermediation organisations (Figure 3.2). Funding institutions work as switches selecting (on) or rejecting (off) ideas for potential innovations. Priorities are placed on intangible infrastructures, human skills, intellectual capital, innovation financing, cooperation and social capital.

In 1994, core concepts of the 'Learning Region' paradigm (collaborative networks, organisational learning, institutional agreements, social capital, triple helix consensus) were adopted by the European Commission, which introduced a new family of policy schemes that took a strategic view of technology and innovation at regional level. The EU launched four types of regional innovation programmes: (1) Regional Innovation and Technology Transfer Infrastructures and Strategies (RITTS); (2) Regional Technology Plans (RTP); (3) Regional Innovation Strategies (RIS); and (4) Programmes of Regional Innovative Actions (PRIA). These initiatives provided co-financing and guidance to regional governments to undertake an assessment of their regional innovation potential, to define and implement strategies that promote innovation in small companies, and experiment with small scale projects. The objective was to create robust regional systems of innovation capable of sustaining and facilitating innovation in small companies in manufacturing and services.

European regional innovation policy has covered the entire European territory with 152 regional innovation strategies in 25 Member States, 145 Programmes of Regional Innovative Actions supporting the implementation of innovation strategies, 13 thematic innovation networks, the Innovating Regions of Europe network, and numerous parallel activities and projects for

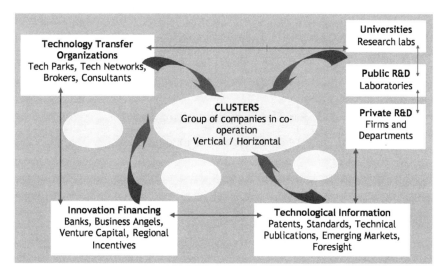

Figure 3.2 Cooperation circuits into regional systems of innovation

innovation funded by the mainstream Regional Operational Programmes. This large-scale, but also strategic, orientation of regional development through innovation implied two assertions: first that regions are key players in the dynamics of innovation; innovation is a challenge for all regions, and all regions should improve their innovation performance; and second that the different regional trajectories apart, it is feasible to adopt a common policy model: the Regional Innovation Strategy (RIS) platform. To comply with the second one, RIS methodology should be generic enough.

Regional Innovation Strategies obey four key methodological principles:

- RIS are based on public–private partnerships and the consensus between academia, business, and government. The private sector and the key regional research and technological innovation players should be closely associated in the development of the strategy and its implementation.
- RIS are demand-led initiatives, focusing on firms' innovation needs, and are bottom-up with a broad involvement of regional actors in their elaboration.
- RIS are action-oriented. At the end of the process new innovation projects in firms and/or new innovation policy schemes should be implemented.
- RIS exploit the European dimension through engaging in inter-regional co-operation and benchmarking of policies and methods.

In parallel, common monitoring of regional innovation performance was developed by the Trend Chart action. The Innovation Scoreboard publishes an annual report that highlights the performance of EU Member States, regions, and industry sectors in the innovation economy. Member States and regions are compared with respect to 26 indicators and a composite index, the Summary Innovation Index (SII). The latter is formed by equal weighting between all indicators and normalisation based on relative to EU-25 data, using rescaling with 0 as lower bound and 1 as upper bound.

Figure 3.3 Innovation policy adapted to regional systems of innovation

Innovation policy inspired by regional systems of innovation leads to five-fold strategic priorities, supporting respectively the regional R&D capability, technology transfer, company cooperation and innovation capability, innovation financing schemes, and information management and diffusion.

Regional innovation strategies and systemic approaches are now at the core of EU regional policy. This is clearly reflected in the new objectives for 2007–2013 (convergence, competitiveness and employment, and European territorial cooperation), in which innovation appears as the top cohesion priority.

Multi-cluster systems

This is the usual form of regional systems of innovation. It arises from the co-existence of many innovative clusters within a region: technology districts in various sectors of industry, Science and Technology Parks with different areas of specialisation, more than one incubator, and other forms of innovation networks. The individual clusters share joint infrastructure, human resources and creativities, business services, collaborations with research bodies, and are funded by the same institutions and mechanisms.

The innovation system that emerges has multiple focuses with independent cores, but is also marked by overlaps and interfaces. A new institutional, overlying innovation system covers the individual technology districts and clusters. It ensures a central governance structure.

In multi-cluster innovation systems one should expect that competition will emerge between the individual cores comprising it. Clusters may compete for human resources, particularly executives, for funding, access to public research, and the liaison interfaces. Furthermore, different cluster development strategies (endogenous–exogenous strategies) result in competition over the choices of public administration and the infrastructure that will be created.

An overlying innovation system and its extended governance are capable of transcending and even bringing together the competitive forces between individual clusters and islands of innovation. Whether the regulation provided is based on a participative approach and ongoing dialogue or on top-down planning and specialisation of the islands of innovation in different technology fields and markets, it is certain that such an arrangement permits cohesive, collaboration-based relationships to emerge and for the area overall to be identified as a highly creative and innovative locale.

Precarious regional systems of innovation

Do all regions have innovation systems? The answer is yes, but under some conditions: that it is possible to identify innovation efforts in products and technologies; that it is possible to identify cooperative networks and relationships that link organisations of R&D, production, and funding; and that there is a rudimentary form of governance at play which attempts to plan and rationalise the way that core elements of the innovation system operate.

68 *Globalisation of innovation and intelligent cities*

In most less developed regions the innovation systems are rather precarious. Despite that, they do exist. Their incomplete form is demonstrated by the lack of key players and all the cooperation circuits found in a fully developed innovation system. From my experience in a series of regional innovation strategies across Europe, what is usually missing are venture capital funds, R&D labs working with companies, and technological intelligence organisations. On the contrary, frequent is the presence of technology mediation organisations and pubic liaison offices and technology transfer centres.

A precarious regional system of innovation may take one of the forms outlined in Figure 3.4. The technology collaboration networks included are 'partial':

- only between research (R&D), technology transfer (TT) and production organisations (C);
- only between information (INFO), funding (FU) and production organisations (C); and
- only between information (INFO), research (R&D) and production organisations (C).

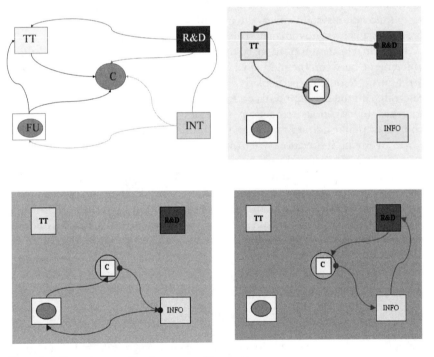

Figure 3.4 Precarious regional systems of innovation
R&D: Research and development – TT: Technology transfer – INT: Information intelligence – C: Companies/Clusters – FU: Funding.

Among the central efforts of European regional innovation strategies (RIS) has been to verify the precarious nature of systems of innovation in less developed regions and to develop institutions/mechanism to complete them.

Regional Innovation Poles

The development of Regional Innovation Poles is based on experience acquired by the EU in the period 1994–2004 concerning the design and operation of regional strategies and systems of innovation. It is a further step in the field of regional innovation strategies that addresses certain weaknesses in those strategies, such as their very wide scope covering all the industry sectors and components of the innovation system, the lack of a global perspective in innovation networking, and the uncoupling of strategy from funding immediately available to implement it.

Both regional innovation strategies and poles focus on the regional system of innovation. They are approaches that adopt a systemic view of innovation as a process of collaboration and integration of R&D, technology transfer, and new product development institutions and skills. In the case of Poles, we are referring to a system of innovation focusing on a small number of industry sectors and cutting edge technologies, with a clear framework for collaboration between research, technology transfer, and entrepreneurialism bodies. The rationale of Regional Innovation Poles is to create an environment favouring innovation, which is characterised by three key features: (1) sectoral targeting; (2) powerful management capabilities; and (3) a direct link between the Pole's strategy and projects to implement it.

Each Regional Innovation Pole attempts to establish a ***sectoral system of innovation***. It may focus on new industrial sectors (telecoms, computers, electronics, new materials, scientific instruments and apparatus), knowledge-intensive services (IT, media, finance, consultancy, medical services) or on traditional industries (food, clothing, furniture, etc.) where they are associated with innovation and new technology activities, and services (tourism, transport, etc.).

> The notion of *sectoral system of innovation and production* complements other concepts within the *innovation system* literature (Edquist,1997) such as *national systems of innovation* delimited by national boundaries and focused on the role of non-firms organisations and institutions (Freeman,1987; Nelson 1993; and Lundvall 1993), *regional/local innovation systems* in which the boundary is the region (Cooke *et al.*,1997) and *technological systems and the distributed innovation system,* in which the focus is mainly on networks of agents for the generation, diffusion and utilisation of technologies and for innovation (Carlsson-Stankiewitz,1995; Hughes,1984; Callon,1992, Andersen-Metcalfe-Tether, 2000). In a sectoral system perspective, it is recognized that national and regional/local boundaries matter to varying

degrees depending upon the specific sector under consideration. Similarly, the sectoral system of innovation approach encompasses and includes the technological system approach, by placing it within the sectoral context and its economic activities processes. And any analysis that takes a particular configuration of technological systems as a 'given' risks overshadowing fundamental processes that serve to define these technological systems.

(Malerba 2002, p. 7)

A Regional Innovation Pole is a network for partnership between organisations in a region, but also an independent organisation that manages regional collaboration in the targeted fields. As a management body it ensures cohesion between research, technology transfer, innovation development, and production activities. It transfers technologies and standards for new products and services via technology dissemination and innovation development schemes. Involvement in the Pole can be direct or indirect. An organisation carrying out Pole projects participates directly, whereas organisations supporting the initiative have indirect involvement.

The French version of innovation poles appeared under the 'Pôles de Compétitivité' (competitiveness clusters). In response to the challenges of the global economy and their impact on the French economy, France in 2004 launched a new innovation strategy focusing on the key factors of industrial competitiveness, particularly R&D and innovation, and the creation of competitiveness clusters.

The definition of a competitiveness cluster in a given region combines three ingredients (Pôles de Compétitivité 2007): businesses, higher education hubs, and research units; and three key factors: partnerships, R&D projects, and international visibility. For instance, the 'SYSTEM@TIC PARIS-REGION' Pole gathers about 200 industrial, academic and institutional members from the Paris-Region who work in partnership on R&D projects related to four target markets: Telecoms, Security and Defence, Automotive and Transportation, System Design and Development Tools (see www.systematic-paris-region.org/index.php?pge=63).

To implement this policy a series of measures was proposed: a three-year budget of €750 million was earmarked to develop competitiveness clusters; the ministries were urged to allocate 25 to 30 per cent of their intervention funds to collaborative innovation projects; businesses participating in collaborative R&D projects could be eligible for exemption from corporate income tax and for lower social security charges; the Caisse des Dépôts et Consignations (CDC), ANVAR (the French innovation agency) and the guarantee funds BDPME/SOFARIS agreed to contribute to this initiative by allocating funds and providing clusters with financial assistance.

Poles were selected by means of an open call for projects. The idea was to select a first series of proposals for the creation of clusters in the area of structuring technologies and industrial activities in which France specialises or has strong potential. Following the first call, 66 Pôles were selected. Six correspond to 'global projects', nine to 'potential global projects' and 52 projects have a national dimension. All projects are developing synergies between the companies, the research institutes, and education institutions within a defined geographical space. Each Pôle is focusing on an industrial sector: high tech sectors (aeronautics, biotechnology and health, multimedia, electronics, software) and more traditional sectors as well (agriculture, automobiles, energy, materials, plastics and chemicals, industrial processes).

The Greek version of Regional Innovation Poles began back in 2002 and the first selection was made in 2006. Five Poles were chosen in respective regions, with a different industrial target for each Pole.

The concept of Pole corresponds to a single-sector cluster or multi-cluster innovation system:

> Regional Innovation Pole (RINPOLE) means a grouping of bodies in the private and wider public sector which seek to bolster the technological and innovation performance of regions in Greece and to increase the competitiveness of the regional economy. Participating bodies should be research and technology bodies, universities, technological educational institutes, businesses, chambers or other agencies engaged in activities related to the said objectives in the region. This grouping is organised around some market or a technological or scientific cluster which seeks to develop and utilise a critical mass of technological and innovative ideas so as to become competitive at national, and in particular, at international level. Current scientific and technology excellence or the ability to develop excellence, as a factor for bolstering the competitive position of the specific geographical area, is a key ingredient in the targets of any RINPOLE.
>
> (GSRT 2005)

Every Innovation Pole bolsters rather than creates a system of innovation in the sector of its choosing. It has to mobilise a significant number of productive and research and technological bodies in the region to generate the critical mass required for an integrated approach to innovation at the targeted sector. The sectoral or technological targeting of each RIP cannot include more than three sectors or three technological fields.

Enlargement III: intelligent global systems of innovation

The rise of the information society, the considerable spread of the Internet, the creation of virtual networks and community digital spaces opened new ways to the functioning of innovation supply chains and systems. Key innovation processes such as technology dissemination, technology learning, product development, marketing and promotion are strongly influenced by digital networks and IT applications. The rich information provided by the Internet opened up new possibilities for strategic intelligence. Online technology market places offered global opportunities for technology transfer. Virtual product development platforms enlarged the networks of cooperation in product development to a global scale.

Intelligent territories, clusters, cities and regions were born out of this convergence. These are systems of innovation combining innovative clusters, technology learning institutions, and digital innovation spaces. They create environments that improve cognitive skills and human capabilities to innovate. They provide superior cognitive capacity and creativity, which is collectively constructed, emerging from the combination of individual cognitive capabilities and information systems that operate within the physical and digital spaces of cities and regions.

Digital/virtual innovation environments add two new dimensions to territorial systems of innovation based on proximity (clusters) and institutions (learning regions):

- A 'global' dimension emerging from the widening of innovation systems on a global scale, enhancing the co-operation of regional actors with global players.
- An 'intelligent' dimension emerging from the use of information and communication technologies, sustaining the global reach of regional innovation systems through advanced information processing, intelligence, and communication capabilities. Digital infrastructure and online innovation services are radically changing the way companies obtain information and assess world markets, the way they transfer and absorb technologies; how they develop new solutions, products, services and organisation.

The creation of intelligent systems of innovation is a three-step process, in which the physical, institutional, and digital dimensions of innovation come together. It is not feasible to create such systems using only digital networks and tools. The digital applications, whether a network, agent or tool, should be integrated into institutional mechanisms for generating innovation (a cluster, innovation centre, incubator, innovation supply chain) and assist people and organisations who work in the context of that institution.

The starting point for making intelligent systems of innovation is the cluster. The cluster represents the simplest form of agglomeration of skills enabling continuous innovation. The complexity of relationships and networks of innovation within a cluster leads to innovation emerging rather than to top-down planning of innovative activity. Existing clusters provide the initial background for intelligent systems of innovation to emerge. Consequently, a starting point is to explore existing productive agglomerations, understand their strengths and weaknesses, and select the most suitable ones. We may call this phase 'awareness' of the initial conditions and opportunities. It is a phase requiring mapping: mapping the internal and external capabilities of existing clusters; mapping networks linking companies, Universities, R&D, financing, technology transfer organisations.

The second phase relates to building up wider innovation networks and support mechanisms around the clusters. Firms in the cluster come into contact with research and innovation development bodies (research units, institutes, consulting firms, new product design centres, financing brokers and product marketers, marketing companies); partnership networks are formed which expand the cluster and form the external innovation environment for its businesses. Within this environment companies will modify their internal routines and production practices. Innovation will be generated via interaction and collaboration. Search, as evolutionary theory claims, is a point of reference followed by technology acquisition and cooperative new product development.

What makes such wider systems of innovation intelligent is the addition of a third phase, where the system of innovation's functions and mechanisms are coupled and extended via digital networks and digital knowledge management tools. In this third phase, the digital space enhances the innovation capacity, but above all the geographical space from which such capacity can be drawn. New innovation tools emerge:

- strategic intelligence, enabling companies to overview their global external environment with the intention of finding information that can be incorporated into management processes and their internal routines;
- online R&D and technology acquisition, defining technologies that should arrive at the company floor and streamlining innovations-to-the-market towards innovations-to-the-company;
- online cooperative innovation, linking new product development teams within the companies with external specialised product development centres at different stages of the new product development process; and
- online global marketing and placement, marketing of products and services and marketing of the cluster itself; development of digital marketing strategies allowing remote producers and customers to be reached.

To date, two different forms of intelligent systems of innovation have appeared. On the one hand, there is a nodal form, dealing with the physico-virtual clusters, which is based on the creation of a digital layer upon clusters, Science and Technology Parks, and incubators. This layer enables the development of online innovation services and the management of cluster-based innovation processes (infrastructure, attraction, technology transfer, spin-off creation) using digital tools and interaction. On the other hand, there is also a more diffused form covering wider areas of cities and regions, based on the creation of virtual innovation spaces by private and public initiatives and the integration of dispersed information systems having different software architectures and content.

Governance of territorial systems of innovation

If we look closer at the recent history of regional innovation in Europe, we observe a gradual evolution of territorial systems of innovation: small business clusters based on trust and cooperation within local communities have evolved towards larger regional systems and are further enlarged due to IT infrastructure and online innovation services. What this trajectory reveals is a movement from simple to more complex territorial systems of innovation. In cluster-type innovation systems an institutional dimension is added, and then a digital one. The lesson from the European experience is that innovation systems start from simple forms, which are gradually enriched with new elements coming from institutions, and information technologies. The corresponding space is under continuous enlargement, from small cores of innovation to wider regional systems, and global supply chains.

In regions with low organisational and technological capability, either in Europe or in developing countries, this route from simple to more complex and from local to global innovation systems is highly relevant. Regional strategies, instead of dealing with all the range of complexities arising from low R&D investments, limited patent activity, low presence of high tech sectors, limited innovation capabilities in the small business sector, etc., which usually characterise innovation backwardness, should focus on igniting this step-by-step building from simple to complex systems of innovation.

The governance of territorial innovation systems should comply with the evolving character of innovation towards continuous innovation and widening systems. Continuous innovation demands an uninterrupted flow of creativity, skills, and knowledge leading to a widening of innovation networks over a global scale in order to assure this uninterrupted flow of skills.

Depending on the territory in question, governance should integrate multiple innovation support actions into a coherent and long-term perspective: from cluster development, to institutional building for R&D and innovation, and the creation of virtual innovation environments. The system of innovation should be developing in physical, institutional, and digital spaces, with hard infrastructure, institutions, and online applications. The aim of this multi-dimensional deployment is to combine the advantages that different forms of territorial innovation systems offer, while taking into account the main challenges to deal with (Table 3.2).

There are a number of issues, however, that this new type of governance of innovation systems should carefully consider and take into account:

- First, identify and manage clusters taking into account the emerging and bottom-up dynamic of inter-firm collaboration. The usual risk is adopting top-down planning procedures not taking into account the members' choices and the complexity of flexible collaborative networks. It is also equally risky to invest highly in hard infrastructure (i.e. in Technology Parks) while neglecting technology transfer and spillovers or to over-emphasise attraction practices leading to devastating inter-regional competition for attraction packages and incentives.

Table 3.2 Advantages and challenges in territorial systems of innovation

Territorial systems of innovation	Advantages	Challenges
Technology districts and clusters	• Direct participation of companies • Well-known and widespread concept	• Planning barriers • Limited geographical scope • High infrastructure cost
Regional systems of innovation – learning regions	• Wider systems compared to clusters • Support from regional policy – long-term intervention • Multi-cluster alliances • Involvement of multiple institutions • University R&D inputs • Public innovation funding	• Need for high institutional thickness • Need for strong public–private partnerships
Global physical-digital innovation systems Intelligent cities and regions	• Add a new dimension to physical and institutional systems of innovation • Tap into global knowledge deposits • Global communication and networks • Round the clock, low cost communication	• Internet diffusion • IT literacy • Complex environments • Global innovation supply chains

- Second, identify the institutional actions that complement cluster strategies. There are three risks in this undertaking: selecting actions that suffer from weak sustainability due to insufficient demand or high cost of maintaining the service they offer; selecting actions that suffer from weak consensus because of limited support from regional stakeholders; and neglecting the implementation framework in place, whether institutional or financial.
- Third, develop carefully selected forms of virtual innovation environments, such as regional intelligence, online innovation management, virtual clusters and communities, digital promotion platforms, corresponding to the needs and capabilities of local users. The usual risk is to create digital services that do not correspond to defined target groups and real communities of users.
- Fourth, create appropriate innovation monitoring and assessment systems. There are many options and solutions for this task: from the Oslo Manual and the EU innovation scoreboard, the Massachusetts innovation economy

index, the UNIDO industrial performance scoreboard, to the UN millennium indicators. Each solution has different advantages and weaknesses and the point is to find a solution enabling the progress of targets and actions to be monitored with the progress made on the performance of the regional system of innovation.

Theory and governance models may define a framework for action and help avoid mistakes and documented impasses. However, the challenge for innovation system governance goes beyond established models. Every innovation system has its own route to follow, valorising a particular mix of resources. At the end of the day, the very meaning of innovation is the capacity to challenge the validity of established models and policy recipes adopted in other regions. This is a central message of systemic open-ended trajectories.

4 Virtual innovation environments

Enriching innovation systems with global networks and users participation

In many respects, the information society and the Internet revolutionised the processes of innovation. The search for knowledge, creativities, and skills were extended all over the world; cooperation among spatially dispersed product development teams was made possible; digital market places opened new global markets. The space of innovation flows became truly global.

To illustrate the effects of digital space upon innovation we will discuss the typology and added value of virtual innovation environments (VIEs). We focus on lessons learnt from the European network VERITE, a thematic network created under the 5th R&D Framework Programme whose objective was to promote tools and technologies for virtual innovation environments.

We start from the relationships between innovation, communities and systems pointing out how major theories of innovation explain the role and contribution of the external environment to innovation. Then we turn to virtual innovation environments, their typology, role, and how they enhance communication, interaction, and problem-solving capabilities. In the final sections we discuss our experiences from the VERITE network, describing the main building blocks of the network, the activities and the tools for internal and external communication, and the impact of VERITE in terms of using virtual innovation tools and spaces.

The wider issue is the relationship between physical, institutional, and digital spaces of innovation. We address three questions with respect to lessons learnt from VERITE: the equivalence of virtual spaces to physical and institutional innovation environments; the added value of virtual innovation environments; and the blending of physical and digital spaces in the innovation process.

Innovation as an environmental condition

Newer theories of innovation attribute an important role to communities and networks that organise and carry on the processes of innovation: interactions within the community, complementary roles and skills, communication channels, functional and spatial bonds, cooperation networks, bridging of separate knowledge fields are all ingredients of participatory, creative processes that result in new technologies and products. The leading role of communities and geographical agglomerations in innovation is acknowledged by most contemporary explanations of how innovation is produced. Out of a rich literature, we can outline three ways of understanding the impact of an external environment on individual creativity, presented by 'brokering', 'tacit knowledge' and 'systemic' explanations of the innovation process.

Brokering theories consider that innovation derives at the intersection of various fields of research and technology. In an extremely important book on how to understand the driving motors of innovation, Hargadon (2003) claims that the mechanism for transforming knowledge into new products is based on the functioning of human communities. Communities of people, synergies and partnership networks set in motion creativity mechanisms that bridge different knowledge fields and produce novel goods. He cites numerous examples that support this view, including key events in the history of twentieth century engineering and applications of electromagnetism and systems theory to industrial production. Knowledge conversion mechanisms are established as communities of people. Contrariwise, the operation of innovation mechanisms ties into the operation of creativity-based communities.

Hargadon argues that innovation is a collaborative process in which knowledge and insights from different fields of science and technology are combined and create something new. A critical factor in achieving a new combination of unrelated knowledge is the human community in which different skills and competences are pooled together.[3] The work of Thomas A. Edison at Menlo Park, he argues, offers valuable insights on the role and interaction between community and innovation. Edison moved to Menlo Park in 1876 where he built a laboratory and put together a team of 14 people doing engineering work for the telegraph, electric light, railroad and mining industries, while conducting their own experiments. In six consecutive years the Menlo Park lab generated over 400 patents and was established as a worldwide invention factory. In Hargadon's words:

> Pursuing a strategy of technology brokering, Edison, bridged old worlds and built new ones around innovations that he saw as result. Much of Edison's work combined existing ideas in new ways; in spite of such humble origins, those inventions revolutionized industries. What set Edison's laboratory apart was not the ability to shut itself off from the rest of the world, to create something from nothing, to think outside the box. Exactly the opposite: it was the ability to *connect* that made the lab so innovative.
> (Hargadon 2003, p. 17)

The same process of creating a community of people that bridge knowledge from disconnected production fields was found at another culminant point of twentieth century industry: the making of Model T at the Ford Motor Company in 1914. In this project, Henry Ford gathered engineers who brought together three manufacturing technologies that had evolved over the last hundred years: (1) interchangeable parts which reduced the dependence on skilled craftsmen, well known to arms-makers from the beginning of the nineteenth century; (2) continuous flow production in the order that tasks occur, already being implemented in the cigarette industry; and (3) the assembly line in which the workers stayed in place and the carcasses moved in front of them, applied at that time by Chicago meatpackers (Hargadon 2003, pp. 40–3). Both cases at Menlo Park and Ford Motors point to the same phenomenon: under genuine leadership an innovative community of people unfolds interconnected individual creativities, generating new products and processes.

In another pre-eminent strand of innovation theory, innovation is conceived as an uncertain, cumulative and path-dependent process, which consumes tacit knowledge and converts it to explicit knowledge. The theory developed by Nonaka and Takeuchi (1995) draws on the Japanese corporate experience in new product development. The authors attach a great deal of significance to the transformation of tacit knowledge into explicit knowledge and describe the enormous organisational effort which is needed for this conversion of atypical and personalised knowledge into explicit, modelled know-how and engineering.

Morgan (2001) takes a step further and argues that tacit knowledge is embedded in individuals and organisational routines that have location-specific dimensions and tend to cluster. Tacit knowledge is spatially 'sticky' and, despite the growth of knowledge management tools, is 'not easily communicated other than through personal interaction in a context of shared experiences'. Clustering becomes inevitable in innovative practices, not in a perspective of minimising transaction costs as many American geographers have claimed, but in order to make the innovative behaviour itself possible; because tacit knowledge, on which innovation is primarily based, is communicable and operational with direct contact only, clusters and other forms of agglomeration become preconditions of innovation.

The essential features of the community, such as interaction, cooperation and networking are positively appraised also in systemic theories of innovation. Systemic theories exercised severe polemic on the so-called 'linear model' in which innovation is conceived as a progressive process starting from research, leading to invention, to product development and then commercialisation. The linear model was criticised firstly because it conceived innovation as a continuum of separate stages rather than as interaction and feedback between different functions and factors, and secondly because of the overemphasis given to R&D with respect to non-R&D factors (Evangelista et al. 1998). The 'systems of innovation' explanation on the contrary stresses the role of innovative communities through clustering or regional institutional networks; the role of interactive relationships between innovative organisations on the demand and supply sides; the role of suppliers and customers; the role of funding agents; the role of the wider technological regimes in which companies operate (Lundvall 1992; Nelson and Winter 1982; Storper 1997).

Innovation springs from externalities, knowledge asymmetries, market imperfections, and institutions that select and manage the flow of knowledge.

Learning, both as acquisition and use of existing knowledge and creation of new knowledge, is the key process. The external environment has a decisive role. Nelson and Winters (1982) explain that genetic processes to innovation are regulated by a 'selection environment' that switches on-off the flow between ideas and products. Nations and regions through their R&D and funding institutions provide this selection and regulatory environment, bridging knowledge, competences and resources from different actors, and screening of ideas and technologies through competent or funding organisations (Nelson and Rosenberg 1993, pp. 3–21).

It seems that a common understanding has been achieved, which attributes a major role to the environment of innovation, pointing to an 'open innovation' paradigm (Chesbrough 2003, pp. 43–62) in which valuable ideas, knowledge and skills come from outside the organisation. Thus innovation is clearly an 'environmental condition' (Komninos 2002, pp. 23–33); it is less an individual achievement than the joint effort of communities of people working together, interacting, and sharing common goals and visions. Empirical studies justify this view. For instance, Sternberg argues that surveys in the regions of Baden, Saxony, and the research triangle Hanover–Brunswick–Gottingen confirm the hypothesis that geographic proximity favours intra-regional linkages between innovative actors, be it companies or research institutions; collective learning becomes easier to develop between actors located within a region than in different regions (Sternberg 1998). The work of Cooke also provides a good many concrete examples of interaction within regional innovation systems and their internal network-based functioning logic (Cooke *et al.* 1997 and 1998; Cooke 2003).

Innovative clusters of companies, networks of knowledge and skills, and regional systems of innovation offer creative environments in which innovative practice emerges. In these environments, innovation occurs because people work together and share and combine individual skills and knowledge into greater experiments and common objectives. Innovation emerges from this interaction of skills, collective intelligence, shared creativity and invention.

Virtual innovation environments

At the end of the twentieth century, human ingenuity added a new dimension to the physical and institutional environments of innovation: a digital or virtual dimension. The rise of the information society and the Internet brought into existence various forms of digital gathering, communication and interaction, which strengthen the contribution of agglomerations and communities on the innovation processes. Henceforth, more complex combinations of physical, institutional, and digital spaces are creating innovative environments.

A series of IT applications, solutions and digital tools is actually used to create virtual innovation environments. Most are based on conventional information technologies and Internet-based communication platforms. The core of solutions focuses less on sophisticated telecommunications than on knowledge management. Services and management tools predominate over bandwidth.

Learning processes taking place in digital spaces complement the creative processes that occur in physical and institutional spaces of innovation. Digital spaces are primarily instrumental spaces. They facilitate the setting up of networks enabling organisations to be engaged in R&D and to deploy their technology capabilities, while the use of digital tools and services is improving problem-solving capabilities and know-how. Yet, the contribution of virtual spaces to innovation, their new relationships to physical and social spaces, which till recently provided the only possible means for making environments inductive to innovation, have yet to be defined. A new literature is given birth in this field.

Schwen and Hara (2003) described a number of communities of practice using sophisticated electronic means and digital applications enabling knowledge sharing and learning. Learning was not on the formal agenda, but it was

Table 4.1 Tools for making virtual innovation environments

Objective	Type of function	Type of online tool
Information	Search/find	Search engines; semantic web tools; meta-search engines (clustering of information); alerts; mashups (content syndication)
	Learn about	Web pages; databases; portals; blogs; wikis; online libraries; online newspapers; online newsletters; rss feeds; digital cities; e-learning; business intelligence
Communication	Communication	Electronic mail; webcasting (podcasting); teleconference; photo and video sharing
	Discussion	Discussion forum; e-communities; social networks (i.e. MySpace, Flickr, Facebook)
	Make a demand; give an order	Electronic exchange; e-auctions; e-commerce; social shopping networks
Problem-solving	Consulting	Virtual brokering; online technology transfer; online R&D; market and technology watch; observatories; customer community relations (i.e. TripAdvisor, VirtualTourist)
	Knowledge processing	Online creativity tools; online mind tools; collective intelligence
	Guiding a process	Digital roadmaps; online innovation management tools; collaborative (decentralised) product development; crowdsourcing; product design tools; virtual engineering; online survey tools; virtual customer

a secondary outcome of becoming knowledgeable while working in a professional field. Identity formation is deeply rooted and tacitly held in professional practice. The explicit goal of the community is to support work practices, but equally important seems to be the formation of identity. Smeds and Alvesalo (2003) studied how tele-presence simulation was used in a geographically dispersed community of practice working within a global company engaged in new product development. The case discussed shows that process simulation enabled negotiation between the local practices and the global design process, beside the limitations of the tele-presence solution. Churchill et al. (2004) examined how digital spaces may mix interactions generated from online virtual communities with offline face-to-face events. They argue that people respond positively to digital spaces presenting faces and other indications of community member identity; are attracted to large central displays giving an overall sense of the content in the display; and are sensitive to the rate of content change on digital display. They go on further by designing digital spaces that blend content from both physical and virtual communities, increasing awareness and information from conferences and meeting events. These surveys show that far from promoting the dissolution of communities of practice, virtual communities reinforce the bond of innovative clusters, creating more complex and effective innovation networks, in which technology spillovers and knowledge exchange are accelerated by institutional networks and digital interaction.

Virtual innovation environments accelerate the entire cycle of innovation from the conception of a new idea to the creation of a new product and the promotion to the market. They appear in multiple forms: web-based intelligence to facilitate the collection and elaboration of information; e-communities which help to exchange best practice and exchange views on alternative solutions; online innovation tools guiding the solution of a problem; web-based platforms and digital cities supporting marketing and product promotion.

e-Intelligence

The start of each innovation effort lies in proper information provision. For any organisation attempting to develop a new product, it is important to know precisely the weaknesses of products already in production, how customers view those products, what competing products exist and what their features are, what the expected consumer behaviour trends are, what opportunities are created by the technologies that are already planned.

Collecting, organising and utilising this information can be made easier and improved with the use of advanced IT tools and by making use of information available from the Internet. Applications that have been developed in the field of business and regional intelligence attempt to make information management easier. They combine tools for targeted information mining, processing and dissemination of information to end users (Figure 4.1).

Many different sources are used when collecting information: indicators to measure an organisation's performance, benchmarking against other organisa-

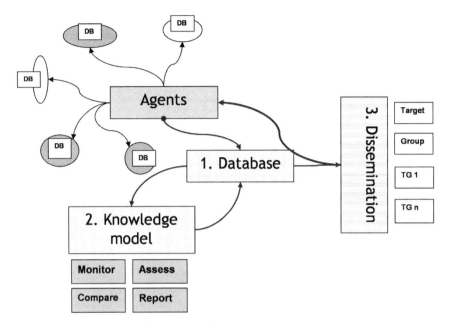

Figure 4.1 Structure of web-based intelligence

tions in the same industrial sector, market trend watching, research and technology outcomes, assessment of future developments via foresight exercises. Agent technologies can also offer significant help in collecting and processing information. Agents denote software-based computer systems that enjoy the properties of autonomy, social ability, reactivity and pro-activeness (Wooldridge and Jennings 1995). Agent-based techniques have been applied to link document frameworks and document management tools, to select the right data, assemble the data, and present the information in the most appropriate way (Knowledge Board 2004).

Processing the information collected from online and offline sources requires an integration or synthesis model. The model determines the thematic structure, special analysis fields, use of complex indicators and systematic comparison and evaluation. Here, information technologies offer less since the tasks involved are exceptionally complex and require human involvement rather than machine intelligence. Several knowledge integration models are necessary if the information recipients operate in different sectors. For example, the information requirements of businesses, the public administration for policy matters, and R&D organisations, all require different information content and synthesis models.

The last stage of information processing relates to disseminating information to selected target groups or individuals. These tasks can be automated to a large extent using web applications, alerts, bulletin boards, and online newsletters.

e-Technology communities

In principle, virtual communities may play a role in innovation similar to physical communities in terms of communication, socialisation, and learning. As in physical communities, the social context of emergence of the community, the evolution of bonds, and intentionality constitute the reference points for describing the characteristics of virtual communities.

> From a social perspective, studies on virtual communities examine the relationship and the social values conveyed by the notion of community. Certain authors question the fact that communities can exist in a virtual mode, since for them the concept of community cannot be dissociated from a common physical space and from history shaped by its members, two elements on which complex social relationships are based. For other authors, virtual life is an established fact and the destiny of human society is from now on dependent on it.
>
> (Henri and Pudelko 2003)

The usual way for a virtual community to emerge is to be created with respect to a network of people sharing the same ideas or objectives. ICTs and digital spaces may help establish the communication channels, and aid information processing and storage. Two situations are possible. A virtual–physical situation in which the virtual space extends the cooperation bonds found in a physical community. In this case, many cooperation relationships remain in the physical space, while part of the communication and learning functions are transferred to the digital space. Some features of the community's physical space may change, but agglomeration factors still remain in operation because the supply chains of goods and materials are partially affected by the digitalisation of information and communication. There is also a virtual–virtual situation, in which the virtual community exists independently from the geographical gathering of people. The virtual space offers the communication channels and determines the identity of the community. In this virtual–virtual situation the social bond is weaker, as is the emergence of intention, and the creation of identity.

Henri and Pudelko (2003), adopting the ideas of Wenger (1998) on learning and communities of practice, categorise virtual communities in four major types with respect to: (1) the emergence of intention; (2) the type of gathering; and (3) the change of intention and gathering over time. These four types are placed into a hierarchical order, from weak to strong social bond and intentionality.

- *A community of interest* is a weak network of people assembled around a topic of common interest. Its members take part in the community to exchange information, to obtain answers to personal questions or problems, to improve their understanding of a subject, to share common passions or to play. Their synergy cannot be assimilated into that of a formal group motivated by a common goal.

- *A goal-oriented community of interest* is not randomly constituted and compares to a task-force or to a project team vested with a specific mandate. It comprises 'expert' individuals, recruited for their competence or their experience, who will share the knowledge and the approaches related to their respective spheres of specialisation. This type of community is created to meet specific needs, to solve a particular problem, to define or carry out a project.
- *A learners' community* is made up of students who may be in the same class, the same institution or geographically dispersed. The learners' community is not permanent because its members are not engaged in a durable way in the activity at the base of its creation. It is born, grows and dies following the rhythm of the stages of an educational programme.
- *A community of practice* develops stronger ties among people who are already part of a given community of practice, engage in the same trade or share the same working conditions. The virtual community of practice emerges from collective activity; it does not constitute an aim in itself but is the result of the involvement of individuals in professional practice activities. For each individual the virtual community of practice represents a means of sharing work practice, of reinforcing professional identity, of enriching or perfecting their daily practice while contributing to the practice of the community.

Virtual communities are also conceived as virtual clusters, 'distinct systems of suppliers, distributors, service providers and clients that use the internetworking technologies as a principal way for co-operating and competing' (Passiante and Secundo 2002). In a virtual cluster, each enterprise adds value by exchanging knowledge with other members. Internet technologies enable increased real-time interaction, e-learning, and technology transfer, which in turn are translated into product improvement and/or a reduction of transaction costs.

Virtual innovation communities work as physical ones: innovation is achieved because of 'collective learning' processes organised within virtual communities or virtual clusters of knowledge. Collective learning involves the 'creation and further development of a base of common or shared knowledge among individuals making up a productive system which allows them to co-ordinate their actions in the resolution of the technological and organisational problems they confront' (Lorenz 1996 cited in Keeble and Wilkinson 1998). Virtual communities generate knowledge that exceeds the capacities, knowledge, and skills of individuals making up the community; creating structures in which there is more than the sum of their components; there is synthesis and synergy.

e-Innovation

This is the most important component of virtual innovation environments. It mainly includes tools and applications for knowledge management and innovation development. The development of online innovation tools requires that complex innovation development processes are deconstructed into individual components. The tools record the steps leading to innovation development and provide guidance to users. In terms of the process they are simulating, they present a logical/functional structure for a topic rather than a morphological virtuality.

These tools can be categorised in various types depending on the internal problem-solving procedure they follow:

- Roadmap tools, which lead to problem resolution step-by-step, providing all the information at each step along with techniques on how to complete it. Roadmaps can provide powerful learning tools which present a problem and oversee the methods that could be used to solve it, accompanied by good practice, bibliographical support, demo and documentation.
- Tools with an input–output structure, where data is input and a result is immediately obtained. This is the case of virtual machines that work along defined internal procedures and data compilation algorithms.
- Collaborative online platforms, which enable a large number of actors to work together and collaboratively solve a problem.

To illustrate what e-innovation tools are about, Figure 4.2 shows key components of the innovation process (intelligence, technology transfer, product development, process innovation) and a series of online tools which facilitates these processes to be implemented. The toolbox of 'Intelligence' combines applications allowing benchmarking and technology and market watch. The toolbox of 'Technology Transfer' includes applications for the dissemination of research, roadmaps to IPR, and training. The toolbox of 'Innovation Financing' combines online tools for drafting business plans, marketing plans, cost-benefit analysis, and an online financing guide. The toolbox of 'Product Innovation' is based on the stage-gate product development model, including tools for ideas generation, evaluation, product design, prototyping, and commercialisation. The toolbox of 'Process Innovation' gathers supply chain management tools, MPR/ERP tools, and information about the management of quality. The application is developed by URENIO and it is available at www.urenio.org/virtual-innovation-environment.html.

Another example is the 'Corporate Innovation Machine' created by Jeffrey Baumgartner, which is an online model for understanding how to implement an effective new idea. The machine comprises several components, all of which must work together for the machine to work properly. When the entire machine does work, it builds ideas, evaluates those ideas and implements the best ideas as new product, service and operational improvements. The machine is available at www.jpb.com/innovation/index.php.

Online collaborative innovation tools range from heavy platforms, like Boeing's platform for the design of the 787 which is occurring simultaneously in Japan, Russia, Italy, and the US, to 'Crowdsourcing' tools, a term coined by *Wired* magazine to describe a business model that depends on work being

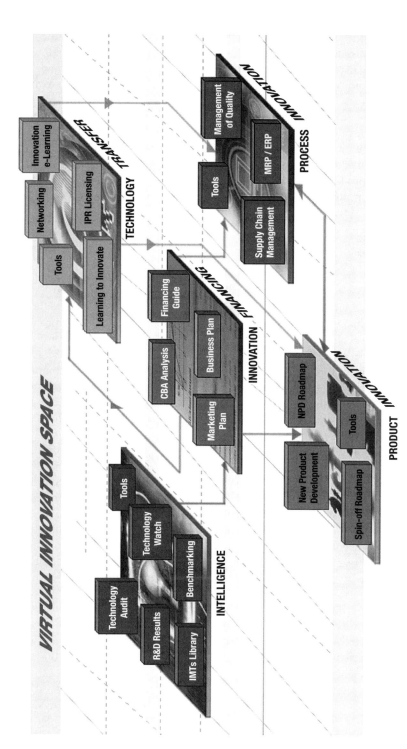

Figure 4.2 Online innovation management tools: overview
Source: www.urenio.org/virtual-innovation-environment.html

done outside the company walls. While outsourcing is typically performed by lower paid professionals, 'crowdsourcing' relies on a combination of volunteers and low-paid amateurs who use their spare time to create content, solve problems, or even do corporate R&D. An example is 'CrowdSpirit', which focuses on harnessing the power of crowds to allow inventors and adaptors to take their products to market. Inventors submit ideas for new or improved electronic products. Contributors vote for the best ideas and define detailed product specifications with manufacturers. Investors invest money in the best products. Testers examine the prototypes and help to fine-tune the products. Ambassadors promote the products to retailers in order to increase the sales coverage of the products. Supporters help by writing or translating manuals, fixing customer product issues, etc. (see http://beta.crowdspirit.com/).

e-Marketplaces

The final step in virtual innovation environments is platforms for marketing products and services (e-shops, e-malls, e-marketplaces, digital cities). After developing new products or services, these platforms offer support in the difficult task of introducing innovations to the marketplace. They enhance product promotion, as well as marketing, advertising, and e-commerce functions.

e-Marketplaces are based on a combination of web, database, multimedia and e-commerce applications. The three major categories are (1) company websites which promote the products and services of a business/organisation; (2) amazon.com-style department stores which agglomerate services and products from many suppliers; and (3) virtual cities which promote products and services from a geographical entity.

Digital cities offer an exceptionally widespread promotion platform for products and services of an area. Along with a presentation of the city, they promote its specific products and services to the market. The architecture of digital cities is multi-tiered: material from the city's physical space, products, infrastructure and services organised in databases, images of the city in 2D and 3D space, navigation applications, e-commerce and online service purchase applications, as well as a user communication interface which combines information, images and applications. Among the latest trend is to combine online services with local broadband networks, wired and/or wireless, which are offered free to the city residents.

The target groups of digital cities are individuals outside the city, visitors, potential investors, and product and service consumers. These individuals come to experience the city in virtual space before actually visiting it in physical space. Most digital cities offer services for digital inclusion, tourism, small business development, and e-government.

The development of digital cities and e-marketplaces all over the world has been spectacular. Good practice is already available by a series of awards and distinctions that many digital cities have received. For instance, the *Age*, an Australian newspaper, has listed the top ten digital cities of the world, which were selected on the basis of broadband speed, cost and availability, wireless Internet access, technology adoption, government support for technology, education and technology culture, and future potential. The top ten are: Seoul, Singapore, Tokyo, Hong Kong, Stockholm, San Francisco, Tallinn, New York, Beijing, and New Songdo City. Different functionalities characterise each city: Broadband in Seoul, free wireless in Singapore, TV over broadband Internet in Hong Kong, education and culture in Stockholm, virtual representation in Tallinn, and integration of all information systems – residential, medical, business in the New Songdo City (*Age* 2007).

Functioning of VIEs

The central question concerning virtual innovation environments (VIEs) relates to how they function. What does this environment offer innovation processes and systems? What innovation functions are made easier or accelerated in digital space? How is this achieved? How are knowledge creation and dissemination mechanisms which foster innovation transformed? What digital tools are available to this end?

As we have seen, the main theories which rationalise how innovation occurs highlight a critical process: knowledge bonds and transformations. Tacit knowledge theories portray innovation as the conversion of atypical and personalised into formal knowledge. In the R&D departments of large businesses there is a continuous transformation of formal and informal knowledge based on socialisation, externalisation, combination, and internalisation of knowledge procedures. Innovation is generated via these knowledge transformations and the interchanges between explicit and tacit knowledge. Brokering theories have shown that innovation derives from the synthesis of various fields of research and technology, and the communities of scientists and engineers within which knowledge and expertise agglomerate. Evolutionary theories also put forward that innovation springs from knowledge asymmetries, market imperfections, and institutions that select and manage the flow of knowledge. Knowledge, both as acquisition and use of existing knowledge and creation of new knowledge, is the key process.

Virtual spaces and innovation digital tools operate on knowledge transactions. As innovation is emerging along with knowledge networks, the virtual environment offers new potential for knowledge generation and management, intervening in the transmission of knowledge, learning, socialisation, knowledge internalisation and externalisation.

Explicit learning: The immediate impact of the virtual environment on the innovation system relates to learning processes which are bolstered by e-learning and e-content applications. e-Learning is the delivery of interactive and digital learning content, aiding knowledge transfer and re-combination of knowledge within communities of practice. e-Learning may develop on different levels, at individual, team, organisation, inter-organisation level, and on different perspectives as well: the cognitive perspective aiming to change the structures of the learning system, the cultural perspective dealing with human behaviour, and the action perspective rooted in experimental learning (Passiante and Secundo 2002). However, the true power of e-learning in the field of innovation is that it transforms tacit to explicit knowledge, bringing to light the innovation processes that occur at the business or cluster level. e-Learning through innovation management tools, for instance, demands an analytical and step-by-step account of the sub-processes of innovation, such as IPR management, new product design, supply chain optimisation, and others.

Skills enhancement: A second aspect of the knowledge innovation system, which is dramatically transformed with the development of virtual spaces, relates to human resource skills. The development of virtual tools and online problem-solving applications is an amazing achievement. Online tools provide employees in the remotest areas with the opportunity to improve their problem-solving skills, and within a short time to acquire new management skills. Online tools differ from e-learning applications because they aid in problem solving without providing analytical understanding of the process. As is the case with all tools, using them does not necessarily entail understanding how they work. In the innovation system, online problem-solving tools are available for all individual steps but above all for market and technology watch, brainstorming, evaluation and product design.

Collaboration: But the field where virtual space has a tremendous impact is partnership networks. Thanks to the constant expansion of innovation networks and systems, already discussed above, the potential for online communication, collaboration and direct response has become a strategic factor in decentralising new product development and production in low cost areas on a global scale.[4] To comply with global production needs and rapidly changing customer attitudes, firms have to continuously monitor customer preferences and changing demand. They have to deploy their networks globally to gather the knowledge they need to support the creation and development of new products. The same knowledge exchange is necessary within supplier–producer relations: in subcontracting where short-term exchanges occur concerning the outsourcing of non-core activities; in technology licensing for exploiting intellectual property; in strategic alliances and flexible agreements to co-develop a new technology or product; in joint ventures where long-term exchanges take place for developing technologies and products. In the context of these multiple and multi-faceted partnership networks, the virtual space offers new functionalities for knowledge exchange and communication in real time. Applications for virtual networking, such as virtual customer, virtual clustering and product development, virtual technology exchange, virtual order placing, virtual follow-up of processes, etc. have greatly amplified the ability of firms to cooperate and innovate. The Internet enables the creation of multiple virtual environments and platforms for collaboration enabling firms located in different parts of the world to work together and tap into customer and supplier knowledge through virtual knowledge brokering and information exchange (Verona et al. 2006). A global window opens to the spatiality of innovation networks and systems.

Reception of VIEs: lessons from the 'VERITE' network

How easy is it to add this virtual dimension in the systems of innovation? And what is the reception of VIEs from the actors of innovation systems? A case study that offers answers and significant lessons to these questions is the network 'VERITE'.

VERITE is a European network focusing on virtual innovation environments; the acronym stands for 'Virtual Environments for Regional Innovation Technologies'. The scope of the network was to enhance learning and good practice related to the creation of virtual environments and the use of information technologies in the field of innovation; in other words, how IT and the Internet can be used to enhance capabilities for strategic intelligence, technology transfer, product, and process innovation at the regional level.

As a case study on virtual innovation environments the significance of VERITE is two-fold: on the one hand, the focus of VERITE has been on the deployment of virtual innovation environments and e-tools for regional innovation; on the other, VERITE was organised as a virtual community exchanging and sharing knowledge on innovation through virtual networks and e-tools.

The network has been set up by 20 European R&D, technology transfer, and regional development organisations dispersed all over the European space; from north to south, from Finland to Greece; and from east to west, from Romania to Spain. Operation of the network gave us the chance to address important questions about the use of virtual spaces in the dissemination of knowledge; the conditions for effective use of digital innovation tools; and the power of virtual communities to substitute physical agglomerations and social networks of innovation.

VERITE started in 2002 as a community of interest but soon evolved into a community of practice. It was one of the fourteen thematic networks of the Innovating Regions in Europe supported by the European Commission in the context of FP5. VERITE was based on the common interest of its members in methods, techniques, tools, and technologies that enable organisations (companies, technology and consulting intermediaries, research institutes, universities) to create and effectively manage innovation. More specific goals were:

- To inform regional authorities, small companies and technology providers in the EU about Innovation Management Technologies (IMTs), especially those applicable to improving regional systems of innovation;
- to help companies and regional authorities reach international expertise concerning IMT implementation;
- to gather and disseminate software and e-tools for implementing IMTs; and
- to encourage the exchange of experience between regions implementing regional innovation strategies about the online management of innovation.

The members of VERITE shared common concerns and professional interests in the field of 'regional innovation' and their cooperation was centred on the creation of environments supporting innovation at the regional level. VERITE was expected to enrich professional practice by sharing and pooling complementary knowledge among its members. A mixture of organisations participated in the network and contributed to it, ranging from technology agencies or institutes (TEKES, Institute Jozef Stefan, etc.), to regional and local authorities (Lorraine, Mantova), to regional development agencies (Centro Sviluppo), to university Labs from Montpellier, Thessaloniki, Thessaly, Wales, and non-profit organisations supporting business interests (Nicosia Chamber of Commerce). Each of these organisations has a different baseline level of knowledge on IMTs and correspondingly different expectations about what the network can offer to its organisation and region.

VERITE was organised as a virtual community. As mentioned, these are communities of people sharing a common interest, goal or practice while a virtual environment supports the community's activity in terms of communication and interaction among members, allowing the spread and socialisation of knowledge inside the community. The network was set up around four building blocks: people and institutions, organisation, focus, and learning.

Table 4.2 Four building blocks of VERITE

NETWORK: People and institutions
20 institutions from eighteen regions of the European Union: university labs, research institutes, regional authorities, regional development agencies, technology intermediary organisations
ORGANISATION: Virtual community
• Face-to-face communication and physical meeting in workshops and conferences • Deployment of virtual space for learning and interaction • Digital interaction among members
FOCUS: e-Tools for managing regional systems of innovation
• Benchmarking and economic intelligence • New product development • Technology transfer • Process innovation • Supply chain management
LEARNING: Acquisition and dissemination
• Thematic conferences and workshops • Online tools: Libraries, forum, e-tools • Regional dissemination of innovation management techniques and e-tools

The people and organisations that formed VERITE and the participants in the network's activities constitute the community of interest/practice. The network partners met in a series of workshops and conferences to discuss the content and implementation of specific innovation management tools and e-technologies. Members also worked together on joint projects in the field of regional innovation.

The network used the Internet in order to support three functions: communication and discussion between members (discussion forum), improvement of knowledge about IMTs (research reports and a database of articles), and improvement of IMT application capability (directory and portal of IMT providers). A virtual space enabling communication, learning, and problem-solving capabilities was created; it is composed of a website that provides information about network activities and stores past information, a directory of innovation management tools and e-tools in particular, a database of articles and research reports on IMTs, and an Internet-based discussion forum.

The focus of the network was on the tools and technologies that enable cities and regions to innovate. The overall field of interest of VERITE concentrates on those IMTs that have a stronger 'network' dimension and can be implemented online. VERITE organised a wide open debate on these issues. Major themes that were investigated concerned:

- *Business and regional intelligence*, including technology watch, economic intelligence, and foresight;
- *benchmarking* at the business and regional levels;
- *technology transfer*, in particular licensing and technology clinics;
- *new product development*, including creativity, product design, testing of prototypes, product life-cycle management; and
- *coordination of the supply chain,* including material requirements planning, enterprise resource planning, supply-chain management, stock management, and outsourcing.

The learning activities of VERITE have developed in three directions. The first direction was workshops and conferences to exchange experience and best practice. Two workshops and five conferences were held during 2002–2003 dealing with the state of the art in selected innovation management technologies and e-innovation tools: Benchmarking in Stuttgart, Germany; New Product Development in Mantova, Italy; Supply Chain Management in Cardiff, UK; Technology Clinics in Helsinki, Finland; and Business Intelligence in Alava, Spain.

- *Company and regional benchmarking conference*: The focus was on the advantages offered to companies and regional authorities by implementing benchmarking initiatives as well as the perspectives of inter-regional cooperation on business performance benchmarking. Considering that existing benchmarking initiatives are based on the calculation of performance indicators

and the definition of best practices at national and international level, VERITE allowed discussion on the basic needs of regions for benchmarking, existing applications and services offered and the perspective of regional synergies.
- *New product development conference*: VERITE was interested in the role of new technologies as a catalyst for product innovation, the role of industrial districts, incubators and technopoles in supporting NPD, and the contribution of funding and policy schemes in supporting NPD. It also examined the regional determinants of NPD and the contribution of the regional environment in the creation or renewal of products; the main sources of ideas for new product development and how it would be possible to exploit and creatively use information provided by remote external sources; tools and methodologies for implementing new product development ideas; management of the product life cycle and the strategies for product replacement or cannibalisation; and applications allowing the sharing of scientific knowledge and expertise along telematics and virtual networks.
- *Supply-chain management conference*: It is widely recognised that supply-chain programmes including manufacturing operations, purchasing and transportation as well as the management of information and financial flows play a critical role in systems of innovation. The focus of VERITE was on how supply-chains can help to achieve the broad objectives of innovation policy; the changing role of suppliers' development strategies, and what regional organisations should be doing to help firms improve their supply chain performance.
- *Technology transfer and 'Technology Clinics' conference*: Among the topics discussed were technology transfer methodologies; 'technology clinics' as applied in Finland; the role of technology providers as coordinators or matchmakers between technology developers and SMEs; and the future of technology transfer. The 'technology clinics' concept is relatively unknown in Europe, though it is a very important tool for technology transfer. The scope of VERITE was to enrich the existing information on this practice and examine the conditions for inter-regional technology cooperation and transfer.
- *Strategic intelligence*: The goal of applying strategic intelligence tools is mainly the need for better information and knowledge; the need to analyse market and technology trends and follow-up of competitors' practices; the need to support and integrate network activities through improving the relationships between research and enterprises; and the need to sustain a service culture in public research institutions.

The second direction was related to the use of virtual spaces for communication and exchange of experiences on innovation management tools and technologies. A toolbox was developed for this purpose, including:

- *A directory of IMT applications* composed of a portal of European providers, a directory of tools, software, and good practice on applications that facilitate the management of innovation;

- *an information dissemination application* based on the construction and use of a database with articles and research reports on IMTs; and
- *an online discussion forum* on issues related to benchmarking, technology transfer, business intelligence, product, and process innovation.

The third direction was dissemination. The activities of VERITE stimulated the dissemination of knowledge and innovation-enabling tools to the partner regions. Much of the dissemination work done by network members took advantage of the contacts and the experience gained during the workshops, conferences, and virtual tools. In Extremadura, Spain, in cooperation with the regional government, FUNDECYT organised meetings with local stakeholders and SMEs, focusing on supply chain management and ERP. External experts were invited to explain these technologies with a question and answer session allowing an exchange of information and needs. In Estonia, ARCHIMEDES supported an initiative to start an IT technology clinic and prepared a one-week knowledge sharing marathon using VERITE forum and online tools. The Nicosia Chamber of Commerce organised a study visit for a dozen entrepreneurs from Cyprus to Belgium. The trip involved presentations from two other VERITE members on innovation management and economic intelligence (Zenit and Regional Council of Lorraine). In Lorraine, VERITE material was used as part of the annual round of training for members of the regional technology development network (ATTELOR). During 2002–2003, the 30 members of the network received some three to four days training on IMTs. In Wales, IMTs have been on the agenda with the Welsh Development Agency leading. Activities were focused more on a sectoral and cluster basis. In Nordrhein-Westphalen, Germany, ZENIT organised a series of activities informing the regional government about supply chain management, including the identification of experts, the organisation of meetings and seminars, and the preparation of a policy using supply chain management as a critical tool for establishing an innovative milieu in NRW.

'VERITE' as a virtual community

VERITE was a community of practice on regional innovation with all the principal characteristics of such communities: the emergence of intention associated with the goal of advancing understanding and use of innovation management technologies; the methods of group creation associated with discussion, conferences, and communication tools; and the temporal evolution and change of both the goals and methods of communication. Given the emphasis on digital communication, VERITE was also a virtual community of practice as online tools developed by its members enabled a constant digital interaction among them.

www.e-innovation.org is the address of the VERITE virtual space, which presents the network's concept, objectives and activities via the Internet. It is the gateway to the network's IT platforms: services and tools portal,

96 *Globalisation of innovation and intelligent cities*

documentation database, and discussion forum. The home page leads to the following sections:

- *Overview*. Presentation of the network's concept, objectives and expected achievements.
- *Members*. Detailed presentation of network's members; logo and profile of each member; contact details, main responsibilities in the network; expertise, and activities in the field of regional innovation.
- *Activities*. This section covers the network meetings. For each of the conferences are included the call for papers, the programme, speakers' CVs, abstracts, and full papers. Presentations from the network's kick-off and technical meetings are also available.
- *Virtual Innovation Community*. This has three sections: Services and Tools, mainly the directory and portal of innovation tools and providers in the

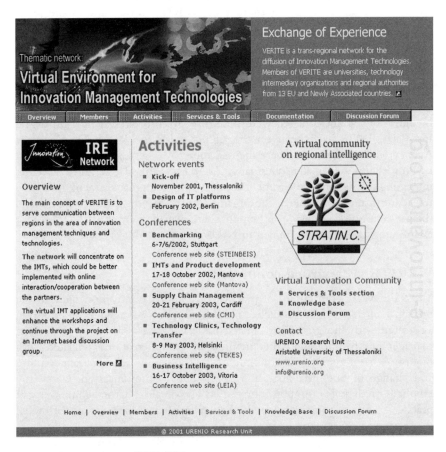

Figure 4.3 Homepage of VERITE
Source: www.e-innovation.org

EU; Documentation Base, where the user may download relevant articles and reports; and the Discussion Forum, for online discussion and exchange of information.

In addition to network members, the virtual space was also aimed at anyone interested in innovation management techniques and technologies, especially benchmarking, new product development, supply chain management, technology clinics and technology transfer, and strategic intelligence.

VERITE mobilised a large audience worldwide from institutions promoting regional innovation policies, innovation experts and technology intermediaries. During the first two years of operation, more than 12,000 unique visitors had access to the virtual space with the majority of visitors coming from EU countries and the US. There was a relatively balanced traffic over the website's whole life, varying from 10,000 to 20,000 hits per month, with the highest frequency in April 2003 when more than 34,000 hits were registered. The most popular section was the Services and Tools section which registered the majority of visitor hits.

Services and tools

This application gives online access to relevant technologies and e-innovation tools that may serve companies, technology intermediaries, and brokers. It is divided into six thematic sections dealing with: (1) Benchmarking and Economic Intelligence; (2) Product Innovation; (3) Technology Transfer and Technology Clinics; (4) Material Requirement Planning (MRP) and Enterprise Resource Planning (ERP); (5) Process Innovation; and (6) Supply Chain Management.

Each section starts with an overview, which includes a description of the main techniques, tools and technologies in the field, and a discussion of objectives, implementation methods, alternative solutions, expected results and benefits. Analytical accounts on each technology cover all aspects of its application; cases where it has been applied, the type of firms and organisations concerned, implementation costs, conditions for implementation, infrastructures required, and European organisations that may contribute to implementation. Then the implementation procedure is described, with phases and steps, tools included in each step, related software, and so on.

Having this overview, the most important technologies and e-tools in each thematic field are listed. The Directory includes about one hundred tools and services. In the field of benchmarking and economic intelligence, tools are about business performance benchmarking, regional benchmarking, and economic intelligence services. In the field of product innovation, tools are about brainstorming, conjoint analysis, creativity development, value analysis, product prototyping and testing, industrial design, product portfolio management, market research, and branding. In the field of technology transfer, tools are about intellectual property management, patent analysis, technology forecasting, and foresight. In process innovation, the tools focus on solutions for lean

Figure 4.4 VERITE services & tools
Source: http://portal.e-innovation.org

manufacturing, energy management and saving, and system automation. In the MRP/ERP section, there are mainly commercial products from European providers or US companies with headquarters and distribution channels in Europe. Then, in the supply chain management section, tools are about systems and services integration, inventory planning, and generic sector-specific tools for process optimisation.

Knowledge base

This virtual library is a database that collects and stores success stories and best practices. The same thematic field of Services and Tools are covered. The database includes around 300 selected papers and reports on benchmarking, business intelligence, technology transfer, product innovation, and supply chain management, offering a good outlook on technologies and innovation tools.

Various sources of information were used to feed the database: papers presented at the VERITE conferences, material collected through the dissemination process from VERITE members and institutions in their regions, and material selected by experts from academic libraries and specialist journals. A discussion forum internal into the database helps to comment on the papers and reports and give advice on issues raised by the users. The database as well as the discussion forum information is open to the general public. However, if a user wants either to enter documents or participate in a discussion of the forum, he/she will have to register before being granted access rights by the database administrator.

Material collection and commentary proved to be relatively weak. Most difficulties came from the reluctance of members to openly criticize the documents and papers included; the language barrier was also an important barrier, since most documents are in English while most regional companies are committed to local language information sources. The application has reported relatively good traffic, with an interesting balance among almost 15 countries with more than 1,000 visits per country without any marketing of the base, apart from some careful placements on Internet search engines.

Discussion forum

Discussion among the VERITE members was facilitated by a discussion forum. It was based on a customised version of an open source application developed by Web Wiz. The first few months of the forum were extremely successful. In May 2003 a hacker destroyed all postings. This was a particularly bad coincidence because the forum had shown strong potential as communication tool. The forum had more than 5,000 visits. Visits were much more frequent than postings. The cause should be searched in the profiling of the discussion and the role of administrator as animator within the virtual community.

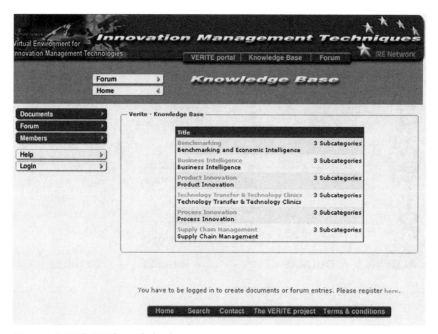

Figure 4.5 VERITE knowledge base
Source: www.e-innovation.org/knowledge_base.html

Learning impact

The learning impact generated by VERITE, both conventional and through its virtual space, was assessed in two consecutive surveys. A first survey took place around the end of the first year of the network's activities with a view to assessing the network member's satisfaction with the activities and online tools (Reid 2003). A second survey took place at the end of the second year with a view to appraising results, satisfaction of the members of the network with the activities, and the likelihood of the on-going sustainability of the network's activities (Reid 2004).

Key questions in both evaluations were, on the one hand, to understand to what extent there was positive impact for members or third parties from the VERITE networking activities; on the other hand, it was crucial to understand the added value of virtual spaces and the perspectives of valorisation of online innovation management tools. For instance, a directory of IMTs can be made easily operational from a technical point of view; however, it only becomes an effective result and gains in value if members of the network begin to use it and update the information.

The 'impact logic' of the network can be characterised in the way shown in the figure below. Activities and inputs of the network are listed on the left-hand side of the figure. Direct outputs of the activities and impact are listed on the right-hand side.

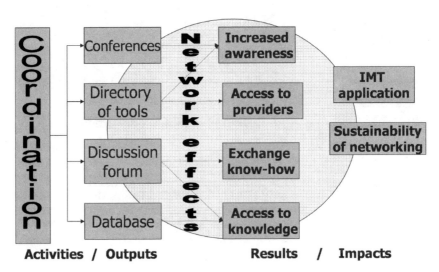

Figure 4.6 Logical impact of VERITE
Source: Reid (2003)

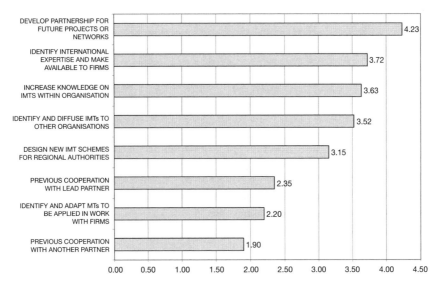

Figure 4.7 Motivation for joining VERITE
Source: Based on Reid (2003)

Motivation for learning

In order to understand the motivation of network members and hence their likely contribution to network activities, each member was asked to indicate the relevance of a number of possible criteria for taking part in the virtual community (Figure 4.7). The predominant incentive for participation in VERITE can be summarised as acquisition of advice or consulting services related to IMTs or an improvement in position in their local 'innovation system' with respect to policy-making. These roles, particularly the first, can be considered as positive to the capacity of the network members to be informed about innovation management, influence policy development, and to act as a channel for raised awareness on IMTs.

A direct conclusion from Figure 4.7 is that the network has apparently not been strongly influenced by past patterns of cooperation because 'previous cooperation with either the lead or another partner' was on average a relatively unimportant reason for joining. At the other extreme, the use of VERITE as a platform for developing future projects or networks activities was a top priority for members. A second conclusion is that the low average score for the option 'Identify and adapt IMTs to be applied in work with firms'(2.28) and the relatively high score for 'Increasing knowledge on IMTs in the organisation'(3.67) tends to confirm that few of the partners were technically expert in IMTs or able to provide IMT services to enterprises. Thus, a learning impact was among the strong incentives for participation.

Direct impact

The direct impact of VERITE was related to the improvement of knowledge on IMTs within the network organisations. Most partners stressed that there was no expectation of a measurable, direct, and immediate impact from participating in VERITE. The benefits of participation were seen as more intangible and often in a more long-term perspective. In many respects, an intangible but real impact of VERITE was to provide 'a vision of where we should be aiming to go in the future' in terms of developing IMT policies. There was some concern as to whether the VERITE approach had either been too generalist (not enough focus on learning on specific tools) or too specific (not a broad enough coverage of IMTs to generate real interest from other regional stakeholders) but this seems to reflect more the approach of the partners concerned with exploiting VERITE than a weakness per se.

This learning advance can be summarised as: Better understanding of state of art in IMTs; improvement in terms of capacity to use IMTs for regional policy making or implementation; involvement in parallel regional innovation activities being undertaken in the region; development of new contacts to be explored in future, and potential for cooperation with international experts.

As a result there was a clear stimulation of new regional schemes enhancing IMTs. Most of the members had not as yet taken significant steps towards developing a new scheme, however a number of initiatives had been taken. Needless to say, most members of the network considered that a standard cycle of development for a new scheme or programme would extend beyond the lifetime of the VERITE network. Despite the regional difficulties faced by a majority of members in developing any new schemes in favour of IMTs in the short-term, at least seven of the network members have made use of the know-how gained through VERITE to push forward or support new initiatives in favour of IMTs:

- In the Basque County, LEIA promoted an initiative on economic intelligence and benchmarking for the wine sector. The Basque government is aware of the initiative and the Secretary for Agriculture is considering funding.
- In Tyrol, VERITE was implemented in parallel with the Regional Innovative Action Programme which started in 2003, of which a fair part is targeted at promoting IMTs in regional companies. A series of training actions had been planned in 2004 which drew on VERITE tools. This involved 20 training projects covering in total about 100 companies. Another training programme has been launched with the local university on IMTs within the context of the MBA course, targeting employees, academics, and intermediaries.
- In Estonia, ARCHIMEDES was involved in supporting two learning projects using VERITE material: the first was a Master of Technology degree based on an e-learning platform. Case material from VERITE has

been used to explain how to develop IMTs. The second was the Tartu University Entrepreneurship and Technology MA which was a classroom-based approach, and so was complementary to the first. In addition, a number of ICT enterprises were interested in developing a technology clinic – but that project is still in its early days.
- In Extremadura, the regional authorities have been highly supportive of VERITE. The main difficulty in take-up of IMTs was considered to be the lack of finance. Hence a proposal to develop a new scheme for funding the use of IMTs in the form of a regional law was being drafted by an expert from FUNDECYT.
- In Nordrhein Westfalen, the ZENIT agency considered that the region was half way to developing a new policy in the context of a shift in innovation objectives to a demand-driven and more systemic approach. The participation of senior officials from the regional Ministry at the Cardiff conference led to a follow-up workshop in Mulheim in July 2003. Experts from VERITE were invited to present experience on supply chain management and IMTs as a policy tool to representatives of various sub-regions, clusters, and regional authorities. The focus of thinking was on how to apply supply chain management to fostering joint learning among groups of firms and the development of the regional 'competence fields' (clusters) already designated.
- In Central Macedonia and Thessaly, URENIO developed a number of online roadmaps in the field of new product development and exploitation of R&D results. The purpose was to create a virtual step-by-step procedure, an online problem solving suite, which enables a problem to be formulated, advanced tools used for solving it, information to be provided on good practice, and the outcome of the problem solving procedure evaluated.

Many members have clearly played an active role in developing spin-off projects related to regional innovation and IMTs. Already there is continuation of co-operation between partner organisations in three projects: STRATINC, managed by the region of Lorraine, which started in 2003 in the context of Interreg IIIC; METAFORESIGHT, managed by URENIO, which started in 2004 in the context of the Regions of Knowledge pilot action; and NPD-NET, managed by the Region of Central Macedonia in the context of Interreg IIIC. All these projects focus on innovation e-tools and exploit the knowledge generated by VERITE, in particular on issues related to business intelligence in innovative clusters, regional intelligence, and new product development. They demonstrate the growing interest in innovation management technologies and e-tools and the critical role played by information and communication technologies as facilitators of innovation.

There were additional knowledge spillovers from the actions to diffuse VERITE results to the partner regions. All partners agreed to draw up a diffusion plan in the form of a local awareness campaign, which involved the design and printing of brochures on IMTs and the organisation of local workshop meetings.

104 *Globalisation of innovation and intelligent cities*

In reality, the network members appear to have adopted various approaches to diffusion taking into account their level of expertise on IMTs and the regional context in terms of development of IMT policies. It seems fair to say that a network like VERITE will have the greatest effect where it is running in parallel with regional initiatives focusing on developing innovation policies or instruments.

A range of partners stressed that a major impact of VERITE had been to improve the profile and image of their organisation with respect to other regional stakeholders. An internationalisation effect (seen now as being part of innovative networks at EU level) could be observed for a number of partners who had been essentially active at regional level before. For others, there was an 'image' effect in terms of being able to offer concrete support or networking opportunities to other regional actors, notably as a complement to other local initiatives (RIS NAC, Regional Innovation Action Programmes, etc.).

Learning and the virtual space

The learning impact of the virtual space and e-tools developed by VERITE was evaluated with a series of questions related to the relevance, efficiency, effectiveness, and contribution to the learning and dissemination objectives of the network. The issue of relevance of the tools being developed was clearly a key topic since the tools were not judged to be useful by a majority of the network members; the networking and learning effects based on digital interaction were likely to be limited. The response of the network members are presented in Figures 4.8 and 4.9.

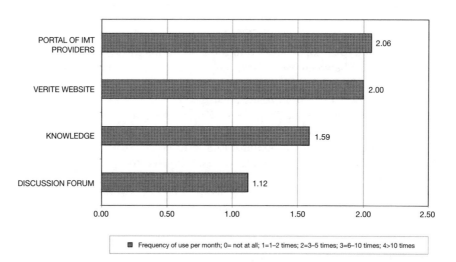

Figure 4.8 Frequency of use of the virtual tools
Source: Based on Reid (2004)

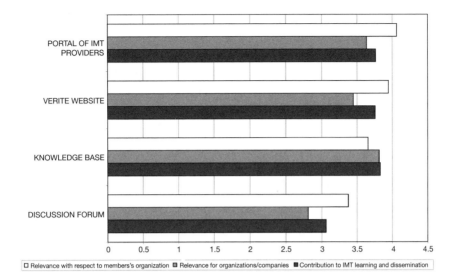

Figure 4.9 Relevance of digital tools of VERITE
Source: Based on Reid (2003)

During the design of the virtual space and online innovation toolkit, certain divergences or at least varying views were expressed. Many members questioned the relevance of the digital tools in terms of a target audience of small companies in the regions, due to language but also technicality of the knowledge to be presented. Serious concerns were raised about who would be updating the IMT's directory and knowledge base with new applications, information, and documents. The usefulness of the discussion forum was also called into question unless it was linked to concrete debates on documents held in the knowledge base.

Somewhat disappointingly with respect to the underlying rationale of VERITE, the response of the network members in terms of the use of the virtual tools was on average rather low. On average, the four types of tools were used only between two and three times a month. The lowest score was obtained by the discussion forum, which in some respects was the most negative aspect since, to be valuable, this tool needs to be used relatively intensively by the partners as a means of continuous knowledge/information exchange. The low score for the knowledge base in terms of use is in contrast with the high score given to the tool in terms of relevance.

Language was a major barrier in this respect. As one member put it:

> While the contents of the tools are extremely relevant, the fact that all information provided is in English is an obstacle to easy dissemination and access by micro and SMEs and entrepreneurs. The knowledge base provided, in our case, can be used only as a source of information for further

local actions. In order for a tool to be user friendly and maximise its impact it has to be available in the final user language. We will translate parts of the Portal into German. SMEs in Tyrol require this.

VERITE was designed with respect to preoccupations about sustainability and continuity. At the end of the first two years of operation, major assets of VERITE were the institutions and people in the network; the virtual space and the online services created; and the accumulated know-how on tools and technologies facilitating regional innovation. The virtual space and tools are key assets of sustainability. Most members considered that it is important to capitalise on the investment made in the virtual tools. However, few members were willing to input more than occasional contributions (documents, experts contact details, etc.) without the incentive of an ongoing cooperation project enabling them to devote resources to maintaining and enriching the applications and information in the portal. On the other hand, a limited number of partners proposed advancing the development of the virtual space by adding e-learning tools, video conferencing or through a more active management of the forum, and stressed that there was a need to develop a real intranet tool to stimulate contacts exclusively among network partners.

Innovation on virtual space

VERITE was clearly a virtual community of practice in the field of innovation, using digital spaces to communicate, learn, and work together than a physical or institutional community: no geographic proximity delimited the network, as members were dispersed over 18 EU regions from the north to south and east to west of Europe; nor were there institutional bonds in operation, as members belonged to different organisations engaged occasionally in cooperation and/or competition. Two years of operation of the virtual community, the participatory design of the virtual space and tools, and the consecutive evaluations on the use of virtual space and tools have provided useful insights on the value of online tools and digital interaction in the adoption and promotion of innovation.

The first question that we are able to address is about the ***role of virtual communities of innovation***. Are they equivalent to local communities or dispersed institutional communities of innovation? Can they substitute the latter? Our reply, with respect to the experience from VERITE, is in the negative.

One main characteristic of a 'real-world' community of innovation is that the social bond is constructed with respect to physical proximity or institutional authority. Physical proximity operates within clusters, while institutional agreements occur in systems of innovation bridging supply and demand factors and regulating the respective technology flows. A virtual community of innovation, on the contrary, is a 'self-defined electronic network of interactive communication organised around a shared interest or purpose' (Castells 1996, cited by Marshall 2000). It occurs in the digital space with respect to digital communication and tools; its existence does not presuppose geographic proximity nor the guiding role of a structuring institution.

Virtual communities are inherently disadvantaged by comparison to physical-world communities. In many cases they are based on a narrow common interest; they tend to be more precarious and live shorter; investments in human and material resources are lower than in physical-world communities. They also are subjected to problems of communication and information transmission that constrain the social bond and cultural breadth in many ways. As Marshall points out:

> Because of their geographical remoteness from each other, the members of a virtual community will initially have little in common except a shared interest. If a community is to persist and thrive, it needs to develop a culture to bind it together. To make this possible, its network must provide both a reliable means of communication and an information transfer capability to support the exchange and interchange necessary for the creation of a common core of ideas. In addition, it would help if the network provided resources for storing some account of the common core so that it could always be available for consultation, for inspection by new members of the community, and so on.
>
> Marshall (2000, p. 405)

The debate on 'physical' versus 'virtual' communities of innovation was extremely well presented by Morgan (2001) who argues that it would be wrong to over-estimate the distance-destroying capacity of information and communication technologies by conflating *spatial* reach with *social* depth. It would also be wrong to assume that because information diffuses easily across territorial borders, that *understanding* does too. Proximity is a highly connecting factor in the tacit and context-dependent processes of learning, knowledge exchange, and innovation. Tacit capabilities are localised and embedded in individuals and organisational routines, and these capabilities have location-specific as well as firm-specific dimensions. Locality, trust, and cooperation matter very much if genuine learning and innovation is to occur.

VERITE strongly validates the above point of view. Information technologies and the Internet do not yet have the power to create or even substitute partially innovative clusters. Online innovation management tools were the strongest elements of the VERITE virtual space, but the software tools we have are still very weak in order to guide innovation processes from distance, and substitute the need for physical presence and communication. Innovation e-tools cannot adequately simulate processes taking place in physical spaces. This might be a major challenge for the future of virtual innovation communities.

The second question we must address is about the **added value of virtual innovation communities**. If they cannot substitute physical-world innovation communities then what is the value of virtual gathering? Though it sounds unorthodox, our answer is that such communities are 'knowledge depositories' and 'problem-solving settings' rather than online communication spaces.

The main assumption behind the creation of VERITE was that innovation depends on the dissemination of new knowledge and knowledge-handling tools. Knowledge related to innovation is specific: how to learn about R&D results and technologies; how to turn an idea into a new product; how to build a spin-off company; how to manage intellectual assets; how to find a new market; how to reduce production and energy costs, and so on. Dissemination of such know-how was the main objective of VERITE. From the setting of the network, all activities and components were designed for the purpose of knowledge reach and diffusion. The network was open to new members; the workshops and conferences too; use of the virtual space was free. The directory and portal of tools were created with the vision of an open library of innovation tools based on open source software. However, this open and cooperative approach was only partly achieved.

Of the three components of the virtual space (communication, learning, tools), the directory of tools and services appeared as the strongest element. Usage of the virtual space by organisations external to the network gives statistics that coincide with the evaluation made by members. The directory/portal of tools attracted the highest level of interest with 7,000 unique viewers, while the database had 42 registered members who all together contributed around 300 articles/documents. The virtual tool that appears to have encountered the least success is the discussion forum compounded by a security failure leading to the tool being hacked in mid-2003 and all postings being lost. Thereafter, the forum was in effect not used.

A virtual community is a collective construct, but the meaning of the gathering is less to exchange views on innovation than to work together and develop tools and processes easily accessible by any member of the community. Innovation tools were also the basis for second-level dissemination inside the member regions, offering the themes for discussion and the necessary documentation and demonstration. If the main problems of innovation are well-documented and known, this is not the case for answers and solutions to them. Virtual communities have a primary role to play in this field.

The third question is about **blending digital and physical spaces of innovation**. The weaknesses of virtual communities may be converted into strengths when they operate with respect to physical or institutional communities. Most disadvantages of virtual communities are removed when communication and knowledge deposits stand on the solid ground of a physical innovative cluster.

The most creative moments of VERITE are to be found in the intersection of direct personal contact and communication and the use of virtual tools. The orientation on 'regional intelligence' as a research theme in projects spinning off from VERITE is a good example of the emergence of a new research orientation from the combination of institutional cooperation and online innovation management tools.

This physical–virtual interface in innovative communities is the essential element of a series of spaces which we characterise as intelligent communities, cities, and regions. The virtual space adds qualities and creative potential by enlarging the horizon of physical clusters and communities, and providing the tools for cooperative work. This added value, however, is subject to the condition of a pre-existing physical or institutional bond.

5 Intelligent cities
The emergence of the concept

Intelligent cities: bridging physical and virtual worlds

The cities always served human societies. They are part of infrastructures and tools we make to confront the challenges of nature and to manage living within societies. However, in contrast to well-designed tools and machines, cities are collective 'tools', whose many features simply emerge rather than being carefully planned. City economies, ecologies, agglomeration characteristics, and innovative capabilities, for instance, can be classed among their emerging features; properties that arise from multiple combinations and innumerable individual choices and practices of the city's population rather than being a result of central control and design.

In the cities and regions of the twenty-first century a radical turn is taking place as information and communication technologies are converging with the rise of innovative agglomerations. What is new here is the turn towards intellectual spaces and tools: instead of constructing spaces that intermediate between nature and mankind's physical condition, as has always happened in the past, a new orientation is arising, attempting to make spaces and collective tools that increase mankind's intellectual capacity and improve the ways we use cities in order to learn, innovate, and reach new frontiers.

> We are on the verge of a new era – for the first time human capital, imagination and intelligence are the keys to success for cities regions and communities. Being competitive in a global market will depend on human ingenuity and innovation more than on natural resources, labour or location,

says the introductory note of INTA's thirtieth World Urban Development Congress that took place in Belfast from 8 to 11 October 2006 (INTA-Annual Congress 2007).

The turn towards intellectual spaces marks a new age in human culture overall. Computers can duplicate many processes associated with the human mind; they can react to stimuli, process information, store and recall information, process symbols and data. Furthermore, they may disappear within the built environment of the city and augment the qualities of physical objects. It is not surprising to find in the post-human literature a deep belief that synthetic intelligence, organic computers and intelligent spaces will deeply challenge the sense of human predominance.

Intelligent clusters, communities, and cities are part of this orientation towards the creation of environments that improve our cognitive skills, our ability to learn and innovate, to foresee, and prevent. They express the challenge of making environments nurturing superior cognitive capacities and creativities, which are collectively constructed and emerge from countless combinations of individual cognitive skills and information systems that operate within the physical and digital spaces of cities.

To date many territories have adopted 'intelligent city' strategies. Public authorities in Singapore, Taipei (China), Spokane (US), Seoul and Songdo (Korea), Cyberjaya and Putrajaya (Malaysia), in many cities of Europe, and in 'smart communities' in the US, have implemented plans to make their cities more 'intelligent'. Each year, the Intelligent Community Forum (2007b) assesses communities, cities and regions from around the entire world and awards those making significant steps towards innovation and broadband development.

In these early attempts two scientific paradigms are setting the scene: cyber cities vs. intelligent communities. Cyber-cities perceive spatial intelligence as a problem of telecommunication infrastructure, digital networking, sensors, intelligent agents, online software applications, and automation in the collection and processing of information; as a pure problem of communication technology and artificial intelligence. At the other end of the spectrum, theories about intelligent communities and intellectual capital for communities understand intelligent cities as a combination of human skills, learning institutions, and digital technologies; integration of these three ingredients enables city intelligence to emerge, and for new city functions, such as strategic intelligence, technology acquisition, and innovation, to materialise.

The term 'intelligent city' (IC) has been used with various meanings. At least five different descriptions of what an intelligent city is can be found in the literature:

- ICs have been initially identified with *virtual reconstructions* of cities, virtual cities. The term has been used interchangeably as an equivalent of 'digital city' covering a wide range of digital representations of the physical space of cities (Droege 1997). We can be certain, however, that the additional communication capabilities offered by a virtual platform or a virtual reconstruction of a city is not adequate to characterise an urban system as 'intelligent'.
- In a metaphorical use of the term, ICs have been seen as a commonplace for various electronic *IT systems and digital applications involved in city operations* and functions. MIMOS, the Malaysian Telecommunications Institute, pointed out that the metaphorical characterisations of the 'Intelligent City' cover many concepts such as the 'invisible city', 'information city', 'wired city', 'telecity', 'knowledge-based city', 'virtual city', 'electronic communities', 'electronic spaces', 'flexicity', 'teletopia', 'cyberville', etc., where each term stresses a specific functional activity.

- Another meaning was given by the World Foundation for Smart Communities, that links digital cities with *smart growth*, a development based on information and commnication technologies. 'A Smart Community is a community that has made a conscious effort to use information technology to transform life and work within its region in significant and fundamental, rather than incremental, ways' (California Institute for Smart Communities 2001).
- ICs were seen as environments with *embedded information and communication technologies* creating interactive spaces that bring computation into the physical world. 'Intelligent environments are spaces in which computation is seamlessly used to enhance ordinary activity. One of the driving forces behind the emerging interest in highly interactive environments is to make computers not only genuinely user-friendly but also essentially invisible to the user' (Steventon and Wright 2006). From this perspective, intelligent cities (or intelligent spaces more generally) refer to physical environments in which information and communication technologies and sensor systems disappear as they become embedded into physical objects and the surroundings we live, travel, and work in. The goal here is to allow computers to take part in activities they were never previously involved in and allow people to interact with computers via gesture, voice, movement, and context. The 'Intelligent Room' is a good illustration of such environments; it is laboratory room that supports computer vision, speech recognition, and movement tracking, based on about 50 distinct intercommunication software agents that run on interconnected computers (Cohen 1997). However, as Bowen-James (1997) and Novak (1997) have pointed out, the critical question is not whether we can build intelligent environments, but how we can use these environments as instruments for distributed problem-solving.
- Intelligent cities were also defined as territories that bring *innovation and ICTs within the same locality*. The Intelligent Community Forum has developed a list of criteria for understanding how communities and regions can gain a competitive edge by combining broadband communications to businesses, government facilities and residences with effective education, training, and innovation in the public and private sectors (ICF 2006).
- Along the same lines, intelligent cities (communities, clusters, regions) were defined as territories of high capacity for learning and innovation sustained by digital spaces and virtual environments of knowledge management and innovation. 'We use the term "intelligent city" to characterise areas which have the ability to support learning, technological development, and innovation procedures on the one hand, with digital spaces, information processing, and knowledge transfer on the other hand' (Komninos 2002, p. 198).

In fact, intelligent cities and regions have all the above mentioned characteristics. They are territories with a high capacity for learning and innovation, which is built-in: (1) the creativity of their population; (2) their institutions for knowledge creation; and (3) their digital infrastructure and services for communication and knowledge management. The distinctive characteristic (and ultimate measure) of intelligent cities is their performance in the field of innovation, because innovation and solving of new problems are unique features of intelligence. Intelligent cities and regions constitute advanced territorial systems of innovation, in which the institutional mechanisms for knowledge creation and application are facilitated by digital spaces and online tools for communication and knowledge management.

The diversity in understanding what intelligent cities are is due to the multiple scientific and technology disciples and social movements that are involved in their creation, namely the movements towards 'cyber-cities', 'smart growth', 'intelligent communities' and 'intelligent environments'. We should stress that all major players shaping intelligent cities promote a form of urban development linked to innovation, smart growth, and digital community spaces. No doubt, there is no formal and commonly accepted definition of what an 'intelligent city' is. Nonetheless, that does not mean that it is not possible to define this concept literally rather than metaphorically in order to characterise an urban system as capable of developing learning skills, creativity, memory, interaction, innovation features, etc., which we identify as features of intelligence when otherwise referring to living organisms. In any case, intelligence does not lie in the 'stones' and the building materials of cities, but in the organised human community and the intelligent tools and infrastructure it disposes.

Examining different attempts to create 'intelligent cities' we find a series of milestones in the movements of 'cyber-cities', 'smart communities', 'intelligent communities', 'innovation environments and living labs', which offered the necessary experimentation and the first examples of the physical–virtual urban systems that we characterise as 'intelligent'.

Cyberspace and cyber-cities

Cyberspace denotes a virtual world built entirely of computers and existing within computers and computer networks around the globe. The initial meaning of the term, introduced by William Gibson in the science fiction novel *Neuromancer,* referred to a dark notion, a dystopian future based on virtual reality, artificial intelligence, and high-tech implants (Gibson 1984; Wikipedia-Cyberspace 2007). Cyberspace was a computer space, but also a 'consensual hallucination experienced daily by billions of legitimate operators in every nation', a 'graphical representation of data abstracted from the banks of every computer in the human system', a world in which everyday life has a strong digital dimension and human intelligence fuses with synthetic intelligence and machine components. In this world, information highways, digital data, and logical circuits were far more important than physical reality and human contact.

In the years that followed, this dark meaning radically changed and actually cyberspace came to be used to describe an emerging universe of virtual spaces

114 *Globalisation of innovation and intelligent cities*

existing within the worldwide computer network, the Internet and the World Wide Web (Graham 2003). Cyber-cities are virtual entities linked to the physical and social environment of cities in two ways: first, they represent this environment with the help of maps, plans, two-and-three dimensional images, and text; second, they help in managing this environment through communication and governance of functions and processes that take place within cities. This second aspect of cyber cities is more compatible with the origin of the term, which comes from 'Cybernetics', a theory of *communication* and *control* that places emphasis on the functional relations between the different parts of a system, and in particular, the transfer of information, *feedback* mechanisms, and *self-organisation* (Heylighen and Joslyn 2001).

Cyberspace and cyber-cities have some unique spatial features that make them extremely valuable for managing the informational aspects of the physical and social environment of cities: physical distance has gone and accessibility is just a few 'clicks' away depending on topological linkages only (Shiode 1997); cyber spaces can be easily modified; digital representations are not delimited by the characteristics of physical space; the production of digital space is extremely rapid and low-cost compared to the production of physical space; digital communication amplifies physical communication and contact within urban communities. Using these features, city planners can create digital constructs, which complement and facilitate activities in the daily life of cities and substitute city functions and spaces. An augmented urban reality is formed.

Discussing the relationships between cities and cyber cities, Pierre Lévy outlined four principles, which should govern these new systems of interactive online communication, and how to integrate or complement cyberspace with normal living space (Lévy 1997):

1 *Analogy in modelling the cyber city*: The 'Digital City' should avoid duplicating institutionalised reality in cyberspace. Copying the cyber-city along the lines of the physical city institutions should not be the overall concept governing the relationships between cyberspace and the city. The new forms of work and interaction, and the instruments of cyberspace enable citizens to get more 'value' and 'profit' out of the distributed intelligence and connected communities in cyberspace rather than simply imitating physical space.

2 *Substitution of urban activities*: New online techniques for collaboration offered in cyberspace no longer require people's physical presence on the spot, making travel unnecessary. The result is less congestion, better traffic circulation, less pollution, easier population spread, and less investment in hard infrastructure. The same positive effects of tele-working may be offered by tele-education applied to universities and higher education, offering interactive, joint educational systems accessible from all territories.

3 *Assimilation of the new networks* into old-type infrastructure existing in the cities (railroads, highways, water, gas and electricity networks, television cable networks and telephone lines). While assimilation is favoured by city

administrations, Lévy objects to this overlapping as he sees the worldwide network of computers and the interactive means of communication not as infrastructure, but as a way of using existing infrastructures. He advocates an active involvement of the user as the key element of cyberspace is not the consumption of information, but the participation in a social process of collective intelligence.
4   *Articulation between the city and the cyber-city*. The social processes in these two spaces are diametrically opposed. The institutions of the territory are rigid and hierarchical, while activities in cyberspace are more transversal and flexible. Articulating the two domains is a process that is far from 'eliminating' territorial forms and 'replacing' them with cyberspace procedures. On the contrary, it is vital to compensate the inertia and rigidity of the territory by making it explicit in cyberspace, and offering expertise, resources and ideas to its solution.

Among the most important functionalities of cyber cities we should also stress the substitution of physical space and infrastructure by digital ones; a process known as *'dematerialisation'*. The classic concept of dematerialisation refers to the reduction over time of the quantity of materials used in industrial products or the energy integrated into these products. It also refers to the absolute or relative reduction in the quantity of materials required for the economic functions of a society. It is a process that clearly concerns the man-made environment, since using less material reduces the volume of waste produced, limits exposure to hazardous substances, preserves the landscape, and even limits the destruction of rare resources and raw materials. Over the last decade, however, the use of digital spaces and the Internet has widened the scope of 'dematerialisation' because of the positive impact of the digital economy on material production and infrastructure. There is abundant evidence on how the digital economy is reducing raw material consumption, the use of packaging materials, and energy consumption. Dematerialisation promoted by the digital economy is not limited solely to industrial production and the introduction of materials and energy saving technologies; it also concerns urban infrastructure, which cities may manage more effectively with the assistance of sensors and online surveillance systems; and traffic as well as digital spaces may diminish the underlying reasons for travel.

It is extremely difficult to make an estimation of existing cyber-cities on the Internet as their number is increasing every day. Multiple cyber-cities agglomerate around each physical city of our world. Besides the number, their variety is extremely limited and they more or less fall into three major categories and their combinations. The first category is comprised of **portal-type cyber-cities**, whose main function is information about city life and activities. The second is comprised of **mirror-type cyber-cities**, two- or three-dimensional city representations, which visualise and present the physical space of cities with static images, three-dimensional panoramics, and/or video. The third category is **tool-type cyber-cities**, which use advanced multimedia to manage and optimise city functions and networks virtually.

Smart communities

Smart communities mark another milestone on the roadmap towards intelligent cities. The initiative for Smart Communities was the first systematic effort to promote cyber-cities and a development of cities based on ICT.

The World Foundation for Smart Communities launched the movement in 1997 in close cooperation with the California Institute at San Diego State University, which drafted the 'Smart Communities Guide'. The mission of the Foundation is to provide a worldwide forum for companies, public administration and academia, disseminating information on the best information technology solutions appropriate for making cyber-cities.

A 'smart community' is a community ranging in size from a neighbourhood to a multi-county region in which public administration, enterprises and residents have understood the opportunities offered by IT and attempt to use those technologies to improve their day-to-day life and work in a significant and efficient manner (Morse 2004).

While traditional cities were built along railroads, waterways, and highways, smart communities are based upon information highways, broadband networks connecting homes, companies, schools and libraries, and applications allowing the sharing and exchange of information and knowledge (Albert 2006). A smart community has four core elements:

- *People/Users*: Each plan to create a smart community starts off from a group of people including the project leader, knowledgeable managers, and motivated users. Their roles are different and getting the community going means defining duties and developing incentives for implementing them.
- *Technical Infrastructure*: The technological foundation of a typical smart community is an information network that links various users in a common purpose. The network includes elements which make communication possible: connections, access points and platforms for the applications. Access points are the terminals which the users have at their disposal, such as personal computers, television, or kiosks. The applications are the uses to which the network's information and resources can be put.
- *Institutional framework*: This includes the community's operating rules, the targets individuals seek to achieve, problem solving mechanisms, and agreement on how infrastructure is to be managed.
- *Applications*: This is at the core of any smart community and determines what any community can do and what the benefits from running the community will be. Applications cover specific fields and key functions within the community. Important fields and community functions covered by applications relates to governance (information on public administration and online administration), entrepreneurship (information and communication on improving business operations, making transactions easier and tele-working), education (collaborative learning, e-learning and online

libraries), and healthcare (dissemination of new medical treatments or the effectiveness of old ones, distance care).

The concepts of 'e-cities' and 'digital cities' are closely related to that of 'smart communities'. An e-city is a community with the necessary ICT networks and systems, applications and software solutions delivering services over local infrastructures to the citizens. Every part of the city is connected to the same infrastructure, running the same protocols, coupled with high speed Internet, e-commerce, and other online services (Tsoukalas and Anthopoulos 2004).

The philosophy underlying the development of smart communities in the US and Canada is inspired to a large extent by 'glocal' principles. City development and prosperity depend less on decisions taken by the central or national government and more on initiatives and guidelines chosen by local leaders. Local leaders should pursue policies for job creation, economic growth, and improving of the quality of life within their region regardless of the policies implemented at the national level. Increased local responsibility is on the rise again in an age when information about markets and growth/development are becoming increasingly global. News, economic and political intelligence, products and services, fly over national borders. Globalisation is transferring decision-making to a lower level, and smart communities offer useful instruments to deal with new challenges in a changing geopolitical context (Eger 1997).

Intelligent communities

The Intelligent Community Forum (ICF) is a parallel, but more advanced effort to understand and promote intelligent cities. It is an initiative of the World Teleport Association seeking to promote the use of information and communication technologies for economic development, in large or small communities, in developed or developing countries.

> "Intelligent Community" is ICF's term for what others have called the wired city, smart community, or e-city. It is the community – whether a town, city, county or region – that views Internet bandwidth as the new essential utility, as vital to economic growth and public welfare as clean water and dependable electricity. Where communities once raced to build seaports, rail depots, airports, and highways to attract businesses and create jobs, many now view broadband communications and information technology as the new keys to prosperity.
> (Intelligent Community Forum 2003)

The Forum conducts research, publishes, and hosts events to spread out this concept. The major achievement of the ICF, however, is the annual 'Intelligent Community' award. Each year ICF selects seven communities from around the world, using five criteria of excellence in information and communication technologies, knowledge and innovation. These are:

- *Broadband infrastructure*, which evaluates the local capacity for digital communication. Broadband is important because companies and institutions are becoming more and more communications-dependent and data-dependent, and broadband is becoming a catalyst for development.
- *Knowledge workforce*, which measures the capacity of the population for qualified work in knowledge-intensive activities. It is not only a matter of university graduates in science and engineering, but also covers knowledge workers on the factory floor, in the research lab, and in the provision of services.
- *Innovation*, which assesses how far communities have gone in creating an innovation-friendly environment that attracts creative people and creative businesses as well as offering access to risk capital that fuels new business growth. Innovation in many cases implies the creation of high tech clusters, but it also means finding a better way to serve customers and deliver services.
- *Digital democracy*, which assesses the government and private sector programmes to overcome the digital divide and ensure that all segments of society benefit from the broadband and information revolution.
- *Marketing*, which assesses the attractiveness of a community and its competitive offerings with respect to other cities and regions. Effective marketing contributes to economic development and leverages the community's broadband, labour and other assets to attract new employers.

The selection of the top seven Intelligent Communities each year starts with nominations made by communities and experts from around the world. The 'Smart21 Communities' is an initial group of cities selected because of their documented strategy for local development and inclusion based on broadband technology. A two-step assessment then follows. An expert Advisory Committee selects the top seven intelligent communities, not because they do extremely well in all the above five criteria, but because each demonstrates excellence in at least one. Later, the best community is chosen with respect to an in-depth analysis conducted by an independent research and consulting firm.

The major contribution made by the ICF through the annual awards is in establishing a strong link between innovation and the information society. The five criteria for selecting the best intelligent communities are allocated into two categories: three are related to innovation and knowledge-based development, and two are related to broadband and access to digital services. It is clear that creating an intelligent community is not a matter of technology only, but of an environment that encourages learning and innovation. Without this environment, the broadband revolution risks worsening social inequality and reducing economic opportunity and political participation rather than creating a new age of prosperity and freedom.

In the cities that have received ICF prizes, two distinct strategies can be discerned. The first is a purely IT strategy focused on the information society which is based around three core elements: broadband, IT training, and e-applications. Characteristic examples are Singapore, Seoul, and Taipei. The second is a strategy focused on the knowledge-based society and economy which combines new economy structures (clusters, high-tech districts, innovation centres, venture capital funds) with broadband infrastructure and e-services. Characteristic examples here are New York, Florida, Glasgow, and Yokosuka.

Intelligent environments of innovation – Living Labs

Examining the approaches taken in North America overall, we can see that intelligent cities result from a convergence of two major trends in modern thinking on the city and urban development: the digital reconstruction of the city on the one hand, in terms of infrastructure and function, and the city as an environment for creativity and innovation on the other.

The European approach to intelligent cities is based on the same mix of digital technology and innovation, but is symmetrically opposed to that of the US. In Europe the starting point for research on intelligent cities and regions was the interest in innovation and territorial innovation systems in particular. The discussion on intelligent cities and regions has primarily emerged in academic writings and EU pilot projects as recognition of the role of social and intellectual capital in innovation and regional wealth creation.

A central effort for sustaining regional systems of innovation and information technologies at the service of cities and regions came from the Innovating Regions in Europe initiatives (RTP, IRIS, RIS, RPIA), which started in the mid-1990s. A series of projects were funded by the European Commission at the intersection of European policies on innovation, the information society, and regional development. In the actual policy orientation there is a clear interest in the convergence of innovation, the information society, and the sustainable development of regions. Three priorities were adopted in the European Commission Communication entitled 'Regions in the New Economy' (EC 2000):

- Regional economies based on knowledge and technological innovation;
- e-EuropeRegio: The information society in the service of regional development;
- Regional identity and sustainable development.

In many of the regions that received public aid in this context, a parallel interest in digital innovation spaces and regional systems of innovation can be noted. Applications include innovation observatories on the Internet, digital dissemination of research results and technology, digital clusters, online technology transfer, digital incubators, etc. Despite not using the terms 'intelligent communities' or 'intelligent clusters' these actions create the main elements of environments that we characterise as 'intelligent', combining advanced digital services in the field of knowledge and innovation management, clusters, innovation infrastructures, and learning institutions.

Recently, another important European initiative in promoting intelligent innovation environments appeared as 'Living Labs'. LLs provide a European platform for collaboration and opening innovative markets in the field of mobile applications and technologies to European citizens, companies, and researchers. It is targeted at cities and regions advancing their telecom infrastructure and digital services in view of becoming significant transaction points for global flows of goods, services, people and ideas. Infrastructures are improved; public policies are adapted to firm specific assets; clusters of competencies are maintained and advanced by applied research and experimental development, education and training. The entire urban environment becomes a 'living laboratory' for prototyping and testing new technology application and new methods of generating and fostering innovation processes in real time.

> A Living Lab is a city area which operates a full-scale urban laboratory and proving ground for inventing, prototyping and marketing new mobile technology applications. A Living Lab includes interactive testing, but is managed as an innovation environment well beyond the test bed functions. As a city-based innovation resource the Living Lab can take advantage of the pools of creative talent, the affluence of socio-cultural diversity, and the unpredictability of inventiveness and imagination in the urban setting.
> (Living Labs Europe 2007)

Ongoing Living Labs initiatives can be found in the cities of Almere (NL), Barcelona (ES), Borås (SE), Budapest (HU), Copenhagen (DK), Hamburg (D), Helsinki (FIN), London (UK), Mataró (ES), Sant Cugat (ES), Sophia-Antipolis (FR), Stuttgart (D), Tallinn (EST), Torino (IT), and Västervik (SE), and this list is rapidly expanding. Living Labs, together with the European Network of Living Labs (ENoLL), which has 51 members (2008) from most EU countries, promote new methodologies for co-creative research and innovation, including open source, open architecture developments, IPR online management, as well as new forms for direct user involvement in the innovation process.

Defining intelligent cities

The above movements, from cyber-cities to Living Labs, allow the concept of intelligent cities to be defined in a more precise manner.

We should note that the concept of 'intelligence' has been always been attributed to the individual, characterising outstanding human mental achievements. Human intelligence has a number of specific characteristics, such as **perception** (allowing to receive and process sensory information to build representations of the world); **communication** (allowing information to be exchanged); **learning and memory** (allowing information to be stored and represented in multiple ways); and **planning and feedback action** (allowing the formulation of goals and the evaluation of progress) (Beckman 2004).

Research on human intelligence clearly links intelligence to innovation, arguing that intelligence is achieving something that has never been done before by the particular individual.

> I think of intelligence as the high-end scenery of neurophysiology – the outcome of many aspects of an individual's brain organisation which bears on doing something one has never done before . . . I like Jean Piaget's emphasis that intelligence is what you use when you don't know what to do. This captures the element of novelty, the coping and groping ability needed when there is no 'right' answer, when business as usual isn't likely to suffice.
>
> (Calvin 1998, pp. 14 and 18)

In the field of artificial intelligence, the concept of intelligence is built on the analogy to human intelligence. Computer intelligence has been assessed with respect to the *Turing Test*. The English mathematician proposed a subjective method to determine whether a computer can manifest the same knowledge skills as a human brain. The test is simple and popularised in science fiction media: an interrogator is isolated, having the task to distinguish between a human and a computer that both answer a series of questions. Machine intelligence exists when it is not possible to distinguish whether the reply to a question has been given by the human or the machine. 'The computer's success at thinking can be quantified by its probability of being misidentified as the human subject' (Turing-Webopedia 2007). However, Fogel argues that a good definition of intelligence should apply to humans and machines equally well, and he defined intelligence as the 'ability of a system to adapt its behaviour to meet its goals in a range of environments' (Fogel 1995, p. 24).

More complex forms of intelligence derive from collective action and collaboration: collective intelligence, swarm intelligence, distributed intelligence of a population or community. These forms of intelligence differ from individual intelligence.

> Collective intelligence . . . is that which overcomes '*groupthink*' and individual *cognitive bias* in order to allow a relatively large number of people to cooperate in one process – leading to reliable action. . . . A less anthropomorphic conception is that a large number of cooperating entities can cooperate so closely as to become indistinguishable from a single organism with a single focus of attention and threshold of action.
>
> (Wikipedia – Collective Intelligence 2007)

Collective intelligence is also meaningful in the field of artificial intelligence. Artificial CI is considered to be an emerging science, based on a large distributed collection of interacting computational processes or multi-agent systems where (i) there is little to no centralised communication or control, and (ii) a world utility function is provided that rates possible histories of the full system

(Wolpert and Tumer 2001). Szuba (2001) proposed a formal model for CI, which assumes an unconscious, random, parallel and distributed computational process run by a social structure.

Looking at intelligence from these perspectives, we can assert that 'intelligent cities' should gather and bring together all three dimensions of human, machine, and social intelligence cited. The concept of the 'intelligent city' should also have the same three dimensions corresponding to human, collective, and artificial intelligence:

- The first dimension relates to people in the city: the intelligence, inventiveness and creativity of the individuals who take part in making the city. This view goes towards what Florida (2002) described as a creative city, gathering the values and desires of the 'new creative class' made by knowledge and talented people, scientists, artists, entrepreneurs, venture capitalists, and other creative people, which have an enormous impact on determining how the workplace is organised, what companies will prosper, whether cities thrive or wither. Cities compete to attract creative people and creativity has become a crucial source of economic growth: to be successful in this emerging creative age, cities and regions must develop, attract and retain those people who generate innovations, develop technology-intensive industries, and create new companies and wealth.
- The second dimension relates to the collective intelligence of a city's population. This derives from institutions of cooperation in knowledge creation and application for solving everyday life problems. In a collection of definitions of collective intelligence provided by Atlee (2005), cooperation and synergy appear as constant elements. Collective intelligence is defined as 'the capacity of human communities to cooperate intellectually in creation, innovation and invention'; 'the collective learning and creative process realised through exchanges of knowledge and intellectual creativity'; 'the capability for a group to organise itself in order to decide upon its own future and control the means to attain it in complex contexts'; 'the sharing of knowledge, know-how and experience in order to generate a higher individual and collective benefit than if they remained alone; 'the co-operation to solve more complex problems than individuals can'; 'the capacity of families, groups, organisations, communities and entire societies to act intelligently as whole, living systems'. Social capital is a key ingredient in achieving collective intelligence. In Putnam's (1995) words, social capital is 'the features of social life – networks, norms, and trust – that enables participants to act together more effectively to pursue shared objectives'. Social capital is needed in most institutions for cooperative innovation, which define the norms for using intellectual property through licensing; creating spin-offs in incubators; getting technology through technology transfer; funding new businesses through venture capital; hosting companies in Technology Parks.

- The third dimension relates to digital spaces and AI available to the city's population to support individual choices and assist communication and cooperation; the virtual spaces that enable cooperation in innovation and the AI applications working in these spaces. What results is intelligent virtual spaces, a public AI, open communication infrastructure and problem-solving tools offering communication in a virtual environment and public digital content and e-tools available to the city's population.

For us the concept of the 'intelligent city' and the plan to implement it, refers to all three aforementioned aspects of the physical, institutional, and digital space of a city. Consequently, speaking literally and no longer metaphorically, the term 'intelligent city' describes a territory (community, district, cluster, city, and region) with four main characteristics:

- a creative population and developed knowledge-intensive activities or clusters of such activities;
- embedded institutions and routines for cooperation in knowledge creation allowing to acquire, adapt, and advance knowledge and know-how;
- a developed broadband infrastructure, digital spaces, e-services, and online knowledge management tools; and
- a proven ability to innovate, manage and resolve problems that appear for the first time, since the capacity to innovate and to manage uncertainty are the critical factors for measuring intelligence.

What emerges from these conditions is a combination of individual, collective, and artificial intelligence, which arises from people, cooperation institutions, and IT infrastructure. It is the intelligence of the community and the intelligent machines it has at its disposal. The intelligence of a city lies in integrating the three tiers outlined: the capabilities of the population, cooperation institutions and digital knowledge management and innovation services.

We would insist that intelligent cities are defined by their innovation or problem-solving capability and the use of IT to improve this capability. Intelligence lies in the problem-solving capability of these communities; it is linked to innovation when a solution to a new problem is attempted; and it is linked to technology transfer when a solution to a known problem is sought. In this sense, intelligence is an internal quality of any territory, of any place, city and region in which innovation processes are facilitated by information and communication technologies. What varies is the degree of intelligence, depending on the people, the system of cooperation, and the digital infrastructure and tools a community offers to its residents.

Three layers of intelligent cities

As we have described it, an intelligent city is a multi-layer territorial system of innovation. It brings together knowledge-intensive activities, cooperation-based institutions for distributed problem-solving, and digital communication spaces to maximise this problem solving capability. It is the most advanced form of territorial system of innovation we have today, a third generation system, following on from clusters and learning regions. It consists of a series of layers, reflecting both the dimensions of intelligence and the development of innovation in physical, institutional, and digital space.

The first layer includes the city's knowledge-intensive activities in manufacturing and services that are usually organised into clusters. The population of the city, knowledge workers, innovative companies – organised in a series of districts – are the fundamental elements upon which intelligent cities are constructed. Proximity in physical space is the critical factor which integrates enterprises, production units, and service providers into a single production and innovation system. Innovation is based on specialisation and cooperation within the city clusters. A critical parameter of this level is the intellectual capital of the city population.

The second layer includes institutional mechanisms for knowledge creation and social cooperation in technology and innovation. Characteristic examples are institutions enhancing R&D, strategic intelligence, venture capital financing, technology transfer, and collaborative new product development. These are mechanisms that promote institutional cooperation within the clusters comprising the city, between different clusters in the city, and between innovation processes taking place on physical and digital space. Critical parameters at this level are institutional thickness in the field of innovation, and the social capital of the city.

The third layer includes digital networks and e-services that make innovation achievement easier. These tangible and intangible infrastructures create virtual innovation environments, based on multimedia tools and interactive technologies, which operate in five pathways towards innovation: market and technology intelligence, technology transfer and IPR, spin-off creation combining R&D results and venture capital financing, collaborative new product development, and process innovation based on cost and transaction-saving technologies. However, the effectiveness of virtual innovation spaces is extremely limited if they are disconnected from creative organisations, people, and clusters.

The endeavour to create intelligent cities is still very much in its early days. To date, most applications are being developed in terms of intelligent networks, clusters and Technology Parks. In these islands of innovation, the physical and institutional innovation system is being enriched with a digital communication and knowledge management dimension, creating an integrated physico-virtual innovation system. There are many signs that such applications will multiply and will cover most city districts. The incentive is strong since the competitiveness of knowledge-based districts and cities increase significantly within intelligent environments, offering various e-facilitators in information, knowledge, cooperation, and innovation.

A glimpse at intelligent cities around the world

Fragments of intelligent cities are emerging all over the world, but still we are very far from the creation of amazing intelligent environments that open minds and radically transform skills and capabilities. This is a weakness both of technology in the field of intelligent environments and of the organisation and integration of technologies with innovation creating activities.

An extremely valuable source of current applications and experimentations in the field is to be found in the Intelligent Community Forum and the cities selected by ICF since 2001 as top intelligent cities (Intelligent Community Forum 2007). 30 cities and regions appear on this list covering a variety of sizes and roles: small cities like Pirai with 23,000 people to multimillion cities like Tianjin with a population of 11 million; global metropoles like New York to small rural communities like Bario, Malaysia; industrial cities and city suburbs (Table 5.1). Among them, awards for Top Intelligent Communities were received by LaGrange, US (2000); New York, US (2001); Calgary, Canada (2002); Glasgow, UK (2004); Mitaka, Japan (2005); Taipei, Taiwan (2006), and Waterloo, Ontario, Canada (2007).

Global metropoles

Global metropoles with their districts and suburbs are frequently among the top seven intelligent communities selected each year: New York; Paris' suburb Issy-les-Moulineaux; Tokyo's suburbs Mitaka and Ichikawa; Seoul and Seoul's Gangnam District; and Singapore. The latter was repeatedly chosen three times among the top seven cities of the year in 2001, 2002, and 2005.

In New York the rise of Silicon Alley in Manhattan sustained the city's specialisation in design, advertising, and publishing. By 1995, the city government began introducing programmes in favour of the new media economy, including 'Digital NY', which provided seed funding to create new high-tech districts. The goal was to extend Silicon Alley into clusters in the five boroughs of NYC.

Issy-les-Moulineaux is a suburb of Paris located near Versailles that made a spectacular effort in the field of technology and ICT. The population has increased considerably because of the location of major IT companies like France Telecom's R&D, Cisco, Hewlett Packard, and Wanadoo. The city population is just 63,000, but employment figures stand at over 70,000, which indicates a strong local productive base and commuting as well. The city government did not spend on IT infrastructure, but focused on services that were widely deployed to citizens. This was a very wise strategy. Private providers offer high-speed Internet services via ADSL-2 or cable modem. More than 70 per cent of residents have Internet access and over 50 per cent are on broadband. A series of new services on training and entertainment are running on the broadband nets, including an annual Worldwide Forum on e-Democracy and the Global Cities Dialogue, a worldwide network of cities promoting an information society free of digital divide and based on sustainable development.

Table 5.1 Top intelligent communities selected by the ICF 2001–2007

	Asia–Australia (10)	Americas (12)	Europe (7)
2001	Bario, Malaysia Singapore (4.42M)	LaGrange, Georgia, US (26K) Nevada, Missouri, US (8.6K) New York, US (8.10M)	Ennis, Ireland (21K) Sunderland, UK (283K)
2002	Bangalore, India (6.00M) Seoul, S. Korea (10.30M)	Calgary, Alberta, CA (900K) Florida, high-tech corridor, US (5.38M)	
2003–2004	Taipei, Taiwan (2.60M) Victoria, Australia (4.70M) Yokosuka, Japan (430K)	Spokane, Washington, US (196K) Western Valley, Nova Scotia, CA (21K)	Glasgow, Scotland, UK (660K)
2005	Mitaka, Japan (173K) Tianjin, China (11.00M)	Pirai, Brazil (23K) Toronto, Ontario, CA (2.48M)	Issy-les-Moulineux, FR (62K)
2006	Gangnam District Seoul (547K) Ichikawa, Japan (466K)	Cleveland, Ohio US (4.10M) Waterloo, Ontario, CA (115K)	Manchester, UK (430K)
2007		Ottawa-Gatineau, Ontario-Quebec, CA (1.15M)	Dundee, Scotland, UK (142K) Tallin, Estonia (401K)

* Each community appears one time only, the year of its first selection. The population in thousands (K) or millions (M) is shown in parentheses.
Source: Based on Intelligent Community Forum (2007)

In Tokyo two areas were identified as intelligent communities. Mitaka, which is a suburb of Tokyo, and Ichikawa, which is located only 20 km away from Tokyo. Mitaka launched initiatives to promote small companies, incubators, venture investments, and high technology businesses. The city hosts a cluster for the design and manufacturing of precision and optical instruments, and has become the worldwide hub for production of animation cartoons, producing about three quarters of all animation seen globally. Computer literacy is extensive, while local broadband connects schools and activity groups in order to introduce citizens to the Internet. The city has a strong education base of academics and researchers. In Ichikawa, the city Cable Network Company introduced broadband services. Total broadband penetration in the city now exceeds 46 per cent, and is higher than the national average. Other initiatives include distance education offered by the Chiba University of Commerce, located in Ichikawa; the pilot system of Japan's Ministry of Education, Culture, Sports, Science and Technology to connect public and school libraries to open up research resources to citizens; and training classes for residents and citizen groups in information and communications technologies. In the Media City, the city's largest project, a Central Library provides access to books, research, music and videos online.

Seoul is considered the world leader in broadband deployment. Broadband infrastructure is a major government priority and service providers offer access to applications such as online gaming, Internet telephony, e-learning, movies-on-demand, finance, shopping, broadcasting, chat, and music. Residents have been inculcated with the 'broadband lifestyle', having incorporated online services into their daily lives. Gangnam is a high-tech district which is home to top South Korean companies and the Teheran Valley information technology cluster. Electronic government has been in use since 1995 with multiple online applications and services for tax payment, cyber civil defence drills, online road control, IP broadcasting for the National Learning Ability Test, and citizen polls. In parallel the district has developed classrooms to deliver computer and Internet training, which were attended by 67 per cent of the population.

In Singapore, the city launched the *Singapore One* initiative providing every citizen with a high-speed Internet connection, training in information technology, and multiple applications including e-commerce, B2B, business information, finding a job, recreation, and many other areas. The vision of the government is 'iN2015' (Intelligent Nation 2015), a strategy guided by the target to make Singapore an 'Intelligent Island' and a global city, powered by ICTs.

Large national cities

A larger number of national centres of industry and services have been identified for applying intelligent community strategies: Sunderland, Glasgow, and Manchester in the UK; Florida and Cleveland, Ohio in the US; Calgary in Canada; Yokosuka in Japan; and Taipei in China.

Glasgow received the award as top intelligent community in 2004 for its protracted effort to ensure broadband development. In the 1980s, the city began reorienting its economy to services, culture and tourism, while at the end of 1990s Scottish Enterprise engaged in projects making Scotland an e-commerce hub. Glasgow is a focal point of national and regional initiatives for e-commerce, and huge investments are being made in high-speed broadband infrastructure, new office space, and the deployment of e-commerce.

Manchester is a well-known industrial, services, and technology centre. With a long industrial tradition, Manchester was the leader in the textile industry and experienced a severe restructuring during the 1980s with job losses and large parts of the population living in deprivation. The city received attention for its efforts to revitalise east Manchester through the *Eastserve* project, a district hit by restructuring. Initially designed as an e-gov application, it included a virtual police station enabling anonymous crime reporting, a home-finder system for public housing, and online job searches and résumé preparation system. In Phase II, *Eastserve* developed a wireless Broadband network linking 1,700 households, six community centres and 14 schools, also offering low-cost recycled computers and training, together with low-interest loans to residents enabling people to participate and understand both the potential of technology and broadband-based economic development.

Sunderland, a city in northeast England with 280,000 inhabitants in the larger metropolitan area, launched a multi-faceted initiative in favour of the knowledge-based economy. Urban renewal projects were at the heart of this policy, including an office park, waterfront, new home for the university, telematics, training, incubators, and e-based businesses, which altogether gave a push to the new economy of the city.

Florida's high-tech corridor was a cooperative effort between universities, the private sector, and local and state government to make Florida a viable place for high-tech industries. It covers 21 counties with 6,800 high-tech companies in various clusters in optics and photonics, medical technology, information technology, aviation, and aerospace. The focus was on bridging broadband infrastructure and services with training, technology education, and workforce development for the new economy.

Cleveland, having recently experienced industrial decline, rising unemployment and racial unrest, made a strong effort for a new future based on IT learning, skills, broadband connections, and online services. Training programmes for low-income, working-age residents were combined with funding to subsidise computer and Internet purchases. *OneCleveland* was created to deploy a metropolitan broadband network and online services for entertainment, health, education, culture, and city planning (One community 2007).

Calgary is the largest city in the province of Alberta, Canada. With a population of around 1 million it has developed a significant telecommunications and wireless manufacturing base with expertise in geomatics and image processing. New infrastructures include innovation centres, fibre-optics, broadband Internet, Infoport, and intangible like venture capital funds, online services and portals, and other facilitators for e-business.

Yokosuka is a port city in Tokyo Bay and home of the new Nippon T&T centre. The city government adopted a vision of development based on ICTs and created the Yokosuka Research Park as an international base for advanced ICTs, high-speed broadband and e-government services. A central hub is the 'Information Frontier City' in which the government, citizens, and local industries work together to implement and expand IT-enhanced government services, to enhance IT in lifestyles, to introduce IT to local industries, and to develop a telecommunications industry.

One of the best illustrations of the parallel deployment of innovation and ICTs is found in the city of Taipei, China, which was selected among the seven best intelligent communities in 2003 and 2006, and received the award as top intelligent community for 2006. The city hosts the country's greatest concentration of high-tech firms. It is a fertile ground for new knowledge-based businesses with 88 technology incubators that launched over 2,000 new businesses during 2005. 45 R&D centres and two major science and technology parks operate in the city, while a third one is under development. The most significant industry continues to be ICTs, a thriving sector in which nearly 400,000 jobs were created during 2004–2006. Taipei is one of the world's top cities for broadband deployment. PCs are in 88 per cent of homes, and an equally high percentage uses ADSL connections. A series of government initiatives are enhancing online services in healthcare, media and banking systems, and e-learning, and direct voice to the administration.

City regions

City regions like Victoria also appear among the top seven intelligent communities singled out by the ICF. The state is in the south-eastern corner of Australia; a small state in terms of area but highly urbanised. Victoria's population is around 5 million, of which more than 70 per cent live in Melbourne, capital and largest city of the region. 'Connecting Victoria' is a government initiative that has set six priorities for regional development: building a learning society, support industries of the future, e-commerce, connecting communities, improving infrastructure and access, and government to be implemented through public–private projects in education, e-commerce, tele-villages, free Internet, and attraction of inward investments in ICTs.

Cities in developing countries

Large cities from developing countries have also shown achievements in intelligent city strategies. Bangalore in India, and Tianjin in China are among them.

Bangalore is one of the world's top centres for information and communication technologies. It hosts a large numbers of multinationals that have set up software development and call centres. The city has a strong local base of universities, engineers, software technology parks, and a commitment to education and training. American and European companies subcontract in Bangalore because of the supply of highly-trained, English-speaking computer engineers, the lower salaries they pay, and the time zone differences that make it possible to run the system 24 hours a day.

Tianjin is located along the *Hai He* River and is a major industrial and technological centre hosting many R&D centres and the Tianjin Economic-Technological Development Area (TEDA), a space where companies can benefit from the city's affluent workforce and resources in terms of supplies, expertise, scientific research and development. In the last ten years, TEDA has experienced fierce development in terms of electronics and communications industries, food, pharmaceuticals, and machinery manufacturing. Digital initiatives of the city have focused on broadband deployment, e-government, on the new high-tech industrial zones. Applications have developed in the fields of e-learning, health and hospital care, tax collection, and port traffic management. On the other hand, TEDA has a local fibre-optical loop offering a wide range of communication services. To promote e-business and e-administration, TEDA is building a domestically advanced urban wideband digital network that makes all categories of wideband applications possible, while in the near future a new high-speed wideband network will enable the fusion of communications, television and computers.

Small communities and clusters

Smaller cities have also frequently been selected as top intelligent communities. In fact, around one-third of the overall communities selected by the ICF as intelligent communities are small.

Ennis is a city of 20,000 people in the West of Ireland, in which a private sector social experiment made this small town the most wired community of Ireland. There were investments in telecom infrastructure, a broadband fibre optics ring, education and training, use of ICTs in local businesses, and a free connection and PC for every citizen. The city is a test bed for interactive services.

LaGrange, Georgia, US, is a rural city of 26,000 people, that constructed four broadband networks serving business, institutions and residents, providing free Internet access, training services, and attracting broadband-based activities. Nevada, Missouri, US also created a telecommunication development corporation, equipped with interactive video classrooms and a multimedia production facility, which became the heart of a vivid local cluster of companies specialising in webpage development, e-commerce, and advertising. In a next step training courses in entrepreneurship and telemedicine were developed.

Western Valley, Nova Scotia, Canada is a city of 21,000 people which, with the support of a Smart Community grant, the development agency, introduced innovation in all critical areas: fibre-optic infrastructure, web-enabling public information and services, information technology courses, IT business incubation centres, and the promotion of IT literacy.

Bario, Malaysia, a modest remote community in the highlands of Borneo, without phones and public electricity, adopted ICTs and the Internet to educate the people of the community and connect local agriculture authorities and to e-commerce a renowned local rice variety.

This epigrammatic review of cities and regions selected as top intelligent communities by the ICF shows that adopting such strategies is not a privilege of wealthy and developed localities only. No doubt, the ICF awards have some symbolic value, but they clearly show that cities in developing countries, smaller, and rural locations may also apply and benefit from intelligent city strategies. There is room for everyone, and this makes these strategies a truly global and universal model.

The other important lesson is that regardless of the size of the city, its position in the urban hierarchy, its location on the world map, its activities and content, intelligent communities are built upon two stable values: a growing set of innovation-led activities, on the one hand, and an expanding network of digital services covering any aspect of the city life, on the other.

Metrics

Is it possible to measure 'city intelligence' or to define a quantitative model for assessing the progress towards intelligent cities and regions? The answer is probably affirmative. Monitoring and measuring are acts of simplification. Numbers have the challenge of clarity, though they conceal many aspects of what we call reality. Any quantitative model is a simplified delimitation of a hyper-complex and infinite reality. However, it is an extremely useful one as well.

Defining metrics in the field of intelligent cities is driven by two principles. First, to compare localities between themselves and learn from the best. Second, to understand the internal dynamics of intelligent cities, define weaknesses, and recognise the effort needed to overcome them. Two methodologies predominate in these attempts: benchmarking and modelling.

Comparing localities that have implemented intelligent cities strategies is the scope of *territorial benchmarking*. Benchmarking has been proven to be a powerful tool of intelligence and the techniques of comparative analysis have spread out in many fields of management and policy development. We may benchmark any type of organisation, institution or geographical entity, provided that we have comparative data from other similar entities. Territorial benchmarking compares and analyses how territorial entities, localities, cities, regions, states perform. It is a rather new form of benchmarking, which looks at the performances of regions and the causes of their performance. It also examines how other territories get something done, how important performance gaps between regions are, which are the territories showing outstanding performances, and which (best) practices are sustaining best performance. Though the specific way that benchmarking is applied in different fields varies very much, the concept and core methodology remains the same. In all cases the process starts with a definition of benchmarking topics, and goes on to select the indicators per topic, followed by data collection, selection of the comparison group, calculation of benchmarks, and interpretation of the results. The scope of the methodology is always the same. We attempt to define the range of variation of performance in any field of activity, the minimum, average, and maximum scores of performance, the distance from the best, and the practices that sustain performances. Identification of best performance and the underlying best practice are the essential pillars of any benchmarking process.

Benchmarking is the simplest way to give meaning to quantitative data and indicators. The work of Florida (2002) on creative cities and class, the Massachusetts Innovation Index, the EU Regional Innovation Scoreboard are benchmarking approaches describing how cities and regions perform in the innovation economy. A good benchmarking example is given also by *City-Vitals*, a benchmarking Table developed by Joseph Cortright at Impresa Consulting. City-Vitals uses a detailed set of statistical measures to understand a city's performance in four key areas: (1) talent (using 5 indicators); (2) innovation (4 indicators); (3) connections (7 indicators); and distinctiveness (4 indicators). Applying this set of 20 indicators, the Table compares and evaluates the performance of the 50 largest metropolitan areas in the US.

North and Kares (2005) proposed a qualitative model for measuring intelligent cities and regions based on the different dimensions of intellectual capital (IC). They argue, for instance, that medieval Ragusa, Bruges in the thirteenth century, and Tuscany in the Renaissance are good examples of intelligent cities that prospered thanks to the use and exploitation of their intellectual capital. They were open and learning regions, with good connectivity and social cohesion. The development of intellectual capital may occur in all activity fields of a city: manufacturing, services, education, research, mobility, communication, health, energy, environmental protection, culture, arts, and leisure. In any of these fields of city activity, they propose ten criteria to assess the development of different forms of intellectual capital: Openness; Vision; Leadership: Cohesion; Self-reflection; Use of ICT; Learning; Connectivity; Initiative; and Experimentation. The authors offer a mapping approach to measure the current position of a territory in terms of intellectual capital development and intelligence. However, their methodology can also be understood as a knowledge or intellectual capital management tool that offers advice on where and how to improve IC in a city or region.

Understanding the internal dynamics of intelligent cities is the scope of *modelling*. Models are schematic descriptions of systems or phenomena that allow for investigation of the properties of the system, the relationships between its elements, the relationships between inputs and outputs, and, estimation of future developments. Modelling intelligent cities means defining the fundamental variables that make a city intelligent, estimating the weight each variable has in the system, and setting some objective function measuring how all elements contribute to the overall performance or goal of the city.

Modelling is more advanced methodology and requires deeper analysis, testing, and understanding. It is mainly about the relationships of the variables that characterise intelligent cities, analysing how different structuring aspects of the city interrelate; and to what extent performance variables of innovation are dependent on the structuring variables of human skills, institutions, and digital infrastructure. The variables are grouped into blocks (skills, R&D, funding, technology transfer, IT networks, e-services, etc.), and the model has to define how blocks are interrelated, and how each block affects the performance of the city in terms of innovation, employment, and wealth.

Table 5.2 Metrics for intelligent cities

Education and skills of the population	Knowledge and innovation institutions	Digital infrastructure and e-services	Innovation performance
1 Population with tertiary-level education (per cent of 25–64 years age class)	1 Number of university students (per cent of total population)	1 City area covered by cable networks (per cent of total area)	1 EPO patent applications (per million of population)
2 Participation in life-long learning (per cent of 25–64 years age class)	2 Number of university staff (per million of population area)	2 City area covered by Wi-Fi networks (per cent of total area)	2 New trade marks (per million of population)
3 New S&E graduates (per cent of population aged 20–9)	3 Total R&D expenditure (per cent of GDP)	3 City area covered by xDSL networks (per cent of total manufacturing enterprises)	3 Innovative enterprises-manufacturing (per cent of all)
4 Researchers in industry and services (per cent of total workforce)	4 Public R&D expenditure (GERD as per cent of GDP)	4 Computers (per million of population)	4 Innovative enterprises-services (per cent of all services)
5 Employment of tertiary-level graduates as per cent of total employment	5 Business R&D expenditure (BERD as per cent of GDP)	5 Internet connections (per million of population) of all enterprises)	5 Enterprises having internal R&D department (per cent)
6 Employment in medium and high-tech manufacturing (per cent of total workforce)	6 Business spending for licensing (per cent of turnover)	6 Broadband connections (per million of population)	6 Sales of new-to-market products (per cent of turnover)
7 Employment in high-tech services (per cent of total workforce)	7 Number of incubators (per million of population)	7 Users of e-gov services (per million of population) of turnover)	7 Sales of new-to-firm not new-to-market products (per cent
	8 Number of S&T Parks (per million of population) enterprises)	8 City enterprises owning a website (per cent of total enterprises)	8 New companies creation (per cent of total enterprises)
	9 Number of Technology Transfer and Innovation Centres (per million of population)	9 City enterprises involved in B2B or B2C (per cent of total enterprises)	9 Exports high-tech products (per cent of total exports)
	10 Exports high-tech services (per cent total exports)		

Both benchmarking and modelling are based on the use of quantitative indicators, which have to cover the entire field of intelligent cities activities and functions. In Table 5.2 we have selected around 35 indicators organised into four blocks, measuring the fundamental dimensions of an intelligent city along the definition provided in the previous section: (1) education and skills of the population; (2) knowledge and innovation institutions; (3) digital infrastructure and e-services; and (4) innovation performance.

Out of these metrics four axis of intelligent city development can be defined. Three of them deal with input factors (skills, knowledge institutions, digital spaces), while the fourth measures outputs (innovation). A 4-dimensions radar chart thus may be defined measuring the progress made in each of the four fundamental dimensions of an intelligent city. There is no need to have all the indicators available to measure the progress made in each dimension.

Both benchmarking and modelling techniques are intended to bring to the surface 'best performance–best practice' relationships highlighting which practice may offer the desired results. We should, however, keep in mind that cities are emerging phenomena and the political, economic, and social factors shaping their performance are beyond the reach and control of a single authority. The social division of labour and market relations make 'best practice–best performance' relationships much more complex and less dependable. It should be borne in mind that intelligent city strategies successfully followed within a particular region may not necessarily generate the relevant results if copied to another region.

Part 2
Building blocks of intelligent cities

6 Strategic economic intelligence
Global watch on markets and technologies

Innovative regions in search of intelligence

A new generation of cities and regions is emerging throughout the world to meet the challenges of innovation-led development and globalisation. They are called 'innovative', 'innovating', 'learning', 'intelligent', and 'regions of knowledge' as well. Their development has gone hand-in-hand with innovation moving out of research labs and the consequent expansion of the spatiality of innovation over the entire world.

A principal feature of innovative regions is their capacity to create environments that favour the turning of knowledge into new products, disseminating information, building organisational learning, integrating skills, and in the end, generating a continuous stream of innovation. Silicon Valley, besides its cyclical ups and downs, has set the standards for innovative regions in the US. However, the same type of innovation-led development, under diverse conditions, takes place all over the world. In northern and southern Europe, from Uusimaa in Finland, Stockholm and Kista in Sweden, Noord-Brabant in the Netherlands, to Sophia-Antipolis and Rhône-Alpes in southern France, Bayern and Baden-Württemberg in southern Germany; all over Japan; in Asia as well, from the Malaysian Cybercities to Singapore, Taipei, China, and Bangalore, India. Regions in developed and developing countries are seeking to benefit from the decentralisation of the high-tech industry and the globalisation of innovation. Innovation-led regional trajectories have rapidly gained ground, constituting the prevailing regional development model at the beginning of the twenty-first century, which more and more regions are trying to copy and implement.

Literature in the fields of economic geography, urban and regional development, and technology management is persistently seeking to explain this type of territorial development, and outline the policies and good practice that set the motors of regional technological innovation in action. Different explanations are offered with respect to flexible specialisation theories on technology districts and innovative clusters introduced by Becattini (1991), Scott (1988b), and Porter (1990); evolutionary theories on learning regions, territorial systems of innovation, organisational learning and tacit knowledge, by Cooke and Morgan (1997), Landabaso (1999), Lundvall (1992), and Storper (1997); and theories linking innovative regions, intellectual capital, and intelligent environments and cities (Bounfour and Edvinsson 2005; Choo 1997; ICF 2001; Komninos 2002; Lim 2001). This theoretical investigation on the one hand explains how the territories of technological innovation were created, the local histories and the trajectories followed in each case, and on the other hand, what their constitutive elements are, and how these elements combine with each other in creating a self-sustaining territorial system of innovation.

Illustrating the factors shaping the 'innovation growth engine' of Silicon Valley, for instance, Cooke highlights five main components:

- Basic research, knowledge generation and application capability of the kind normally found centred on advanced private research or cutting edge public research laboratories. In Silicon Valley, exemplars are Xerox Palo Alto Research Centre (PARC) and Stanford University, PARC being less visible in the 1990s than earlier, and venture capital more visible in search and selection of innovation than before.
- Venture capital is crucial as the means by which ideas that have been screened and selected are given a chance to fly as commercial products or services. Business angels are key in the initial stage; venture funding for the second stage and stock market flotation. However, it is clear that finance is only part of the story and that the hands on management skills that equity investors bring to firms in which they have invested is at least as important. This extends to cluster-building activity where portfolio firms are advised on local inter-trading, for example.
- Law firms are important as gatekeepers, advising firms on appropriate investors, counselors assisting entrepreneurs in accessing other services, and sources of contacts for many things ranging from recruitment to contract manufacturing. Many law firms practice relatively little formal law with technology businesses. Moreover, they often take payment in equity rather than fees. Therefore as 'knowledgeable attorneys' they constitute a second source of external business know-how.
- Specialist consultants in business and technological services including management accountants rather than simple auditing accountancy services, head hunting services and specialist engineering, software, new media, and regulatory advisers or property development services, including specialised public provision.
- A local value chain of firms that can, for example, engage in contract manufacturing, design and fabrication, and various fairly prosaic supplies like logistics, or exhibition organisation and specialised catering services.

(Cooke 2003, p. 403)

The above description of factors shaping a leading innovative region coincides with what we have defined, in a more abstract way, as critical components of a regional system of innovation (Kafkalas and Komninos 1999). We have argued that innovative regions spatially concentrate industrial clusters, research institutes, technology transfer agencies, funding organisations, and information infrastructure, and work as integrators. Integration takes place between the separate components of the regional innovation process (R&D, innovation finance, technology transfer, new product development, and marketing); but integration also takes place between the physical, institutional, and digital spaces over which the innovation processes occur. The distinctive characteristic of innovative territories is that they turn research and scientific knowledge into new products with the involvement of innovation funding, technology transfer, and product development expertise external to the organisation concerned. No doubt, these externalities presuppose an advanced social division of labour in the field of innovation. This seems to be the dominant trend. Increasing 'externalisation' of innovation from company research labs makes outsourcing mainstream practice for large and small firms in order to acquire the critical inputs of funding (through venture capital), technology (through technology transfer), and inter-firm collaboration (through supply chains and digital networks). Innovative companies and organisations, networks among distributed competences, externalities in the innovation processes, integration among physical, institutional and digital innovation spaces, and new-type innovation governance, are key concepts for understanding the functioning of innovative regions and territories.

Everything revolves around knowledge networks. Recent theoretical research has drawn attention to the interactive and collective character of knowledge that generates the innovative capability of companies. Nonaka and Takeuchi (1995) introduced the concept of organisational learning to describe knowledge generation that takes place within a community of interactions. Edquist *et al.* (2001) extended the field of learning interactions from inside the organisation to cooperation networks among organisations. Organisational learning amplifies the knowledge created by individuals and crystallises it into the structure of the organisation. This sharing of knowledge takes a variety of forms involving the acquisition of existing knowledge from public organisations, R&D centres and universities; licensing from other companies; or less formal types of exchange in technological co-operation networks. It also takes the form of co-operative new knowledge creation within R&D consortia of companies, universities and technology intermediaries. In the same direction, Keeble *et al.* (1998) have shown how 'collective learning', a concept developed by Camagni and Lorenz, may contribute to the innovative behaviour of technology clusters. Collective learning describes the capacity of a social environment to facilitate innovative behaviour by the firms that are members of this milieu. This type of shared knowledge enables establishing a common language for talking about technological and organisational problems, for effectively cooperating on a technological project, and for managing hierarchical relations and responsibilities among different occupations assuring the consistency of collective decision making.

Among the foundations of this system of knowledge is **strategic information**, assuring a continual stream of data about markets, technologies, competitors, and emerging trends. Information is spread throughout the innovation system by formal and informal communication channels providing knowledge management tools to increase problem-solving capabilities (Antonelli 2000). This knowledge helps regions to innovate continuously and overcome successful innovation waves. For instance, a major challenge that most innovative territories are facing today concerns changing markets and technologies: keeping up-to-date with ongoing trends in R&D and technology; the pace that markets mature in semiconductors, personal computers, servers, corporate software; the rise of high-tech markets related to smart phones, digital television, web services, and wireless communications; the significance of ongoing investments in innovative products and technologies, such as utility computing, chip sensors, sensor networks, plastic electronics, and the wireless net; and the new sectors related to life and bio-sciences that promise the greatest growth of any industry since computers.

Companies located in innovative regions have to constantly update information and learning in order to remain innovative and competitive. They do it by combining internal and external resources (European Commission 2001b). The two dimensions (internal and external) are reflected in business intelligence vs. cluster/regional intelligence practices. Business intelligence is developed internally by companies to monitor information related to products, markets, and technologies in which they are active. Cluster and regional intelligence, on the contrary, is set-up by third party organisations (regional authorities, business associations, universities, technology intermediaries, etc.) to provide information about wider trends in production, markets, and technologies.

For business, cluster, and regional intelligence, information technologies and the Internet have become primary means. Information systems are fed by the Internet and the dot-com revolution, allowing worldwide co-operation on the exchange of information. Today, as the fast collaborative-net takes off, it is sparking new ways of using the Web 2.0 and getting the best available information. However, it is not only about connectivity. Digital services referring to R&D, technology transfer, and technology funding, allow for the acquisition of best practice and increase the problem-solving capabilities in all organisations, whether large or small, in core or peripheral regions.

Strategic economic intelligence: business, cluster, and regional

The actual interest in economic intelligence is fuelled by a series of ongoing trends that strengthen the role of knowledge in the external business environment, such as the growing outsourcing of innovation as companies contract sophisticated engineering, design, and research and development services from outside suppliers; the location of R&D facilities and global innovation footprints in emerging economies, India and China in particular; the change in R&D practices along the 'fourth generation R&D' in which lead customers and other stakeholders participate in new product development, testing prototypes, system architectures, and dominant designs (Miller and Morris 1999); the 'open innovation' mentality that is based on a landscape of abundant knowledge not restricted to the company's internal know-how, but profiting from the wealth of available external knowledge, R&D, ideas, and intellectual property (Chesbrough 2003).

In this context, information is becoming a strategic asset in all forms of innovation: new product development, strategic partnerships, selection of contractors, finding available technologies, and entering into new markets as well. Companies, clusters, and regional authorities are developing strategies to manage this vital resource. However, little has been written on how they do it; how they deploy intelligence, and strategic intelligence in particular.

A key concept in mastering information is 'intelligence' with its various denominations, 'strategic', 'economic', 'competitive', 'innovation', 'business', 'cluster', 'regional', etc., which denote stand-alone or distributed information networks, software and systems allowing continuous update and learning about markets, competitors, products and technologies, and the wider socio-economic environment. It is rather difficult to set the demarcation lines between the different labels and terms for intelligence; with all forms emphasising the organised and systemic collection, analysis, and dissemination of information for business and development purposes. To be clear, we use two milestones of reference. First, different forms of intelligence (strategic, economic, competitive, innovation, etc.) denote the thematic content and focus of the practice; each form has its suitable rationale and perspective, though the differences between them are rather minor. Second, different levels of intelligence (business, cluster, sector, regional, etc.) denote the agency that undertakes the practice; here the differences are major because the information needs of a company differ substantially from the needs of a cluster or region. It is easy to realise that multiple combinations may appear from matching 'perspective' and 'agency', leading to equivalent forms of 'intelligence'.

Thus economic intelligence belongs to the wider family of strategic, competitive, market, and innovation intelligence. 'Economic Intelligence, concerns the set of concepts, methods and tools which unify all the coordinated actions of research, acquisition, treatment, storage and diffusion of information, relevant to individual or clustered enterprises and organisations in the framework of a strategy' (CETISME 2002, p. 18). As innovation rises on the company agenda, economic and innovation intelligence turn into strategic assets, influencing major decisions regarding the future of the company and its competitive strategy.

Business intelligence

Business intelligence is the basis for discussing any form of economic intelligence. It is defined as an activity to overview the internal and external environment of a company, with the intention of finding information that can be incorporated into its management processes. The term was invented by Howard Dresner in 1989. It is an organised procedure in the service of the company's strategic management, aiming to improve its competitiveness via the collection, processing and dissemination of information useful for controlling its environment. This informed decision-making uses specific tools, mobilises employees, as well as internal and external networks (Herbaux and Chotin 2002). Business intelligence does not include any illegal or underground activity; rather it is a systematic method of getting information, which is exploited for a business purpose, and to that extent a deep gap separates it from its caricature industrial espionage (Simovits and Forsberg 1997). Business intelligence is also conceived as a strategic approach to systematically targeting, tracking, communicating and transforming relevant 'weak signs' into actionable information on which strategic decision-making is based. 'Weak signs' are anticipatory, uncertain, ambiguous, and fragmented pieces of information, thus subject to interpretation and multiple purpose meanings (Rouibah and Ould 2002).

Business intelligence is mainly a company activity. The focus may be on strategic issues, customers and markets, competition, innovation, etc. depending on the needs of the implementing company. It has evolved out of traditional decision-support systems which gradually incorporated in-house databases (1985), data warehousing (1995), customer relationship management (2000), integrated business intelligence applications (2003), and semantic web watch tools (2004). From this evolution, it has the potential to deliver enormous payback to the company, but demands unprecedented integration of information about customers, competition, market conditions, vendors, products, and the entire supply chain (ARTE 2003).

Business intelligence is facilitated by software tools and a number of fundamental inventions concerning the ways of dealing with data. Vendors offer a series of tools and techniques for data analysis, trends description, and evaluation. Typical BI technologies include reporting, online analytical processing (OLAP), key performance indicators scorecards or dashboards, relational database servers, data warehousing, CRM, reporting and query tools, analysis and exploration tools, data visualisation, data mining, web mining, modelling, alert and notification systems. BI vendors provide tools and platforms enabling the delivery of information to decision makers. Global leaders are large enterprise application vendors (i.e. Oracle, Microsoft, SAP) and medium-sized companies (Cognos, Business Objects, Microstrategy).

Recently solutions have focused on the semantic exploitation of data by means of computational intelligence technologies and adaptive business intelligent applications (Back 2002). Assigning meaning to data, delivering knowledge from data, and deriving optimal decision support are key activities for all business fields, from R&D, to technical design, production, quality control, and supply chain management. Adaptive business intelligence integrates *data mining*, using algorithms capable of discovering new or unknown facts from a dataset of information gathered into a relational database system, and *optimisation* based on input-output models (Selby 2002). On the other hand, valuable information about external business factors is readily available on the Web, and Web farming is an approach gaining ground for business intelligence (Pawar and Sharda 1997).

Information assistants and information retrieval tools have been developed for this purpose, which act on behalf of the user: launch the query by using an analysis of documents based on a semantic network defined by the system, and visualise the results according to data presentation or data exploration techniques (Reiterer *et al.* 2000). An interesting collection and appraisal of the latest software solutions is given by Johnson (2006) which bridge competitive intelligence and innovation, considering that the former is much about substitutes, emerging new entrants, ongoing technologies and business models to address the needs of customers. The software presented covers ten applications based on Web 2.0 and provided by Aquity, Chipher, Coemergence, Comintell, Digimind, Novintel, QL2, Strategy, Traction, and Wincite.

Cluster intelligence

At the other side of business intelligence are cluster and regional intelligence. They may be defined as informational nexuses linking the actors of a territory (Bertacchini and Dou 2001; Dou 2000; Raison 1998; Quazzotti *et al.* 1999). It is a network allowing 'an observation strategy towards the competitors, the markets and the environment. These practices lead to an economic intelligence approach, which, when applied to the territory, is called territorial intelligence' (Bertacchini and Dou 2001).

Thus cluster and regional intelligence are distributed forms of intelligence organised along networks of information and cooperation among actors located in proximity to each other. It focuses on the external environment of the company, though it may also include elements of internal information for comparison and benchmarking purposes. As distributed organisational intelligence, it takes a step forward from traditional economic intelligence tools (watch, business intelligence, assessment) corresponding to the need of policy-makers to engage in localised economic intelligence customised to their own needs (Kuhlmann *et al.* 1999). However, what mainly characterises cluster and regional intelligence is the fact that they are organised by third party organisations; their rationality or scope is not bound to a single company or organisation, but to the welfare of an industry cluster, a territory, locality or administrative region.

Cluster and regional intelligence are 'collective' or collaborative. They represent territory-based forms of intelligence enabling a relatively large number of people or organisations to cooperate in a process leading to the definition or solution of a problem. The term 'collective intelligence' relates to an extensive body of knowledge concerned with several subjects such as distributed cognition, distributed knowledge systems, collective learning, connected or networked intelligence, augmented intelligence, etc. These terms describe human communities, organisations and cultures exhibiting 'mind-like' properties, such as perceiving, learning, acting, and problem-solving. Collective intelligence characterises a large number of cooperating entities that work together so closely as to become indistinguishable from a single organism with a single focus of attention and threshold of action (The Collective Intelligence Lab 2007).

Table 6.1 Business, cluster and regional intelligence

	Focus	Responsible organisation	Type of information	Methods	Software
Business intelligence	• Company strategy, including competition, markets and technologies	• Company • Consultant	• Customers • Products • Technologies • Prices • Competitors • Markets • Stock exchange	• Customers watch • Web watch • Relational databases • Data mining • Balance scorecards • Modelling • Optimisation	• Business intelligence applications • Web farming • Stand-alone applications
Cluster intelligence	• Sector trends • Cluster development	• Business association • Cluster manager • Sectoral association • Business network	• Sector • Innovation • Technologies • Markets and opportunities • Funding • Competition	• Benchmarking • Statistical analyses • Web watch • Data presentation/ exploration techniques	• Sectoral watch applications • Network-based applications
Regional intelligence	• Region • Regional development • Regional future	• Regional authority • Development agency • Development alliance	• Growth • Population • Employment • Skills • Investments • Innovation • Social exclusion • Environment • Digital divide	• Benchmarking • Statistical analyses • GIS	• Regional databases • Visualisation tools • Network-based applications

According to Lévy 'collective intelligence' is a social project of varied intelligence, distributed, unceasingly developed, and coordinated in real-time (Lévy 1994). This definition is built upon four axioms: *It is distributed*: in the sense that nobody knows everything, everyone knows something; knowledge is in a community and not in a transcendent entity which would organise its distribution near the company. *It is unceasingly developed*: Lévy insists on the concept of human qualities; each member of a community is carrying a richness which would ensure a place and a contribution in collective intelligence. *It is coordinated in real-time*: the reference here is to cyberspace, to the semantic web in particular, a tool supporting collective intelligence and allowing communication between media on a large scale. Finally, it *leads to an effective mobilisation of competences* as collective intelligence is not a theoretical or philosophical concept, but it enables effective social organisation based on competences, knowledge, and wisdom (Caillard 2007).

In perspective, Lévy argues, the semantic web will open new horizons to collective intelligence. It will allow the creation of a virtual space where the hyperlinks do not point to documents (texts or images) but concepts. In the model he elaborated, this semantic space will be represented by a virtual architecture, a kind of 'abstract city' on several relevant dimensions of representation. This city mirroring collective intelligence will shelter six 'districts' corresponding to mental representations referring to (1) competences; (2) intentions; (3) 'declaratory', 'procedural' and 'existential' knowledge; (4) social networks; (5) technical networks; and (6) social reality. Each district will shelter 'semantic zones', while each zone will be defined in the language of collective intelligence. The informational objects (sciences, arts, skills, institutions, documents, messages, people, and equipment) will be represented as beings which connect the various parts of the city while transporting resources from one zone to the other. By visiting the city, one will thus discover the structure of the relationships between the semantic zones, i.e. the structure of the collective intelligence considered, at the level of a document, a company, a city, a country, covering every aspect of information that circulates on the web (Lévy 2007).

Taking these orientations into account, we would define cluster/regional intelligence as a territorial information system with five characteristics:

- it is a *localised network* of distributed informational modules;
- it is developed by *third party organisations* for the welfare of a cluster or territory;
- it uses *human and artificial intelligence* in the collection, processing, and dissemination of information;
- it communicates via the *Internet*; and
- it is *integrated* so effectively that its constituent parties become indistinguishable for the external user.

Important work on cluster intelligence is taking place in Lorraine, France. The region, with a population of 2.3 million, is located in the western part of

France and exhibits a strong presence of traditional industrial sectors, textile and wood, though 13 technology support centres are gradually forging a change towards new industries and services in pharmaceuticals and aeronautics. Two applications related to cluster intelligence have been developed: DECiLOR and EPINETTE (Veille 2002).

DECiLOR is an application for technological and economic intelligence set up by the Regional Council of Lorraine. It is the heaviest European investment in regional intelligence with €5 million spent over three years. It is aimed at companies in Lorraine in the sectors of timber, logistics, metal works, and pharmaceuticals. The aim is to provide companies with the essential means to make use of economic intelligence: personalised information, search methodologies, and personalised watch corresponding to company environment. To this end, DECiLOR offers information of any type, validated, qualified, and classified in a database for use by companies in Lorraine, while employing methodologies adapted to the constraints of small and medium companies. It is based on a specialist team making up the back office cell in charge of project control, which feeds the sectoral watch centres.

EPINETTE was developed by CRITT-Bois, a technology centre for the wood industry, offering more focused technological intelligence for the needs of the wood cluster. It structures information elaborated by CRITT-Bois, which conducts approximately 150 studies and answers about 200 information requests regarding norms, patents, products and companies, per year. Information is organised in different sectors based on services provided to clients, including:

- a technological survey, which provides information on suppliers, products, and companies in the wood sector; research labs and research results; industry standards and patents; articles in the professional press, and technical documents and files;
- an online audit, with modules allowing audits to be performed in different fields of the business activity;
- a subcontracting search, allowing companies to offer products and technologies to European partners and find partnership in Europe within the context of Innovation Relay Centres;
- a search for products and services, using key words and taking back related companies, patents, research labs, press articles, and technology offers; and
- a documentation base, provided in cooperation with ENSTIB, the national school of technologies and industries of wood in the province of Epinal.

The interface is user-friendly with well-defined and clear information taxonomy; there are constant information renewal and search capabilities. The strengths of the application are in the bonds developed between intelligence, technology expertise and end-users, linking EPINETTE as an information processing hub with a dedicated technological centre (CRITT-Bois) and the wood cluster of Lorraine.

Regional intelligence

To date, applications of regional intelligence have been limited and their structure is rather simple. We have identified a number of institutions (observatories, regional administrations, documentation centres), which collect and disseminate information in organised ways focused on regional audiences. Most are linked to regional administrations and their operating costs are covered by public funds.

An organised attempt to develop regional intelligence is found in the Regional Observatories in the UK regions of the East Midlands, east England, southeast England, the southwest, the northwest, and Yorkshire. These observatories more or less follow the same model: they are based on a network of regional actors, collect statistical data, primarily cover the public aspects of the economic and social life in the region, and disseminate information via the web. They use advanced web applications, combining databases, GIS, automated interfaces, and provide information on many fields, sectors, and activities of the region.

Established in 1999, the *East Midlands Observatory* is a network of organisations with an interest and involvement in information and research. The purpose of the Observatory is to provide the primary regional framework for collecting and sharing high quality, balanced and relevant economic, environmental, social and spatial information and research. Main activities include surveying and researching selected topics related to industries, skills, and regional economic trends, information monitoring and dissemination through a website on which one can find information and statistics about the East Midlands, discover research done by Observatory partners, learn about other organisations' research projects, and link to websites with useful information. Target audiences are the partner organisations, public sector organisations, local businesses, local and regional trade organisations, potential inward investors, educational institutions, and citizens (East of England Observatory 2007).

The *East of England Observatory* provides an information gateway to this region. It is aimed at people and organisations interested in discovering more about the social, economic and environmental development of the east of England (Bedfordshire, Cambridgeshire, Essex, Hertfordshire, Norfolk and Suffolk). Main areas of activity include research (regional census, social exclusion, policy analysis); a review of the regional economic strategy completed in 1999 with a full review every three years; follow-up of indicators on business registration, workforce, R&D, business location, gross value added and employment, manufacturing investment, and productivity. Geographical information systems facilitate mapping regional social and economic performance (East Midlands Observatory 2007).

Yorkshire Futures is the Regional Intelligence Network for Yorkshire and the Humber region, providing information and intelligence about the region, with the aim of improving decision making and better preparing for the future. The vision is to set up an influential and objective network, ensuring that all regional policy decisions are based on robust, reliable and timely information and intelligence, contributing to making Yorkshire and Humber a world-class region. Its main functions are to provide quicker, fuller and more accurate data; to conduct forward looking research to prepare the region for future events and trends; to undertake policy analysis improving decision making, benchmarking and good practice dissemination. Information is organised into about 20 thematic areas, among which are strategies and policy, business competitiveness, workforce and skills, environment, business, futures. Three monthly briefings provide updates about economic trends, policy developments, and EU policies. A novel characteristic is the *Knowledge Rich Programme* (KRICH) through which regional businesses can access vital information before their competitors. Developed by Yorkshire Forward, KRICH is a business information service that provides online access to advice and expertise on innovation, research and new technology development; equipment, facilities and services to support research, testing, new product and process development; intellectual property rights and licensing opportunities; the latest technological, scientific, legislation and management developments; practical guidance on improving business competitiveness and profitability through innovation; and an online forum for sharing experience and learning best practice (Yorkshire Futures 2007).

The *Regional Intelligence Unit* (2007) offers organisations in the northwest region access to key intelligence. The online information2intelligence system enables them to find the information needed with greater ease, efficiency, and speed. The Unit supports the Regional Intelligence Network (RIN), a network of researchers and practitioners from all interested areas in the region, which facilitates the flow of data, information, and best practice. The Unit undertakes a process of identifying intelligence gaps in the region, in partnership and consultation with regional players who are members of the RIN. Using Geographical Information Systems that allow datasets to be analysed and displayed spatially, the Unit undertakes analysis on various economic datasets, and provides a series of regional statistics.

These observatories offer a standard level of regional intelligence, shaped by the target groups they focus on. With the exemption of KRICH, data comes from socio-economic surveys and statistics; that limits target groups to academia and the public administration that are the usual recipients of such data. The scope of information is long-term planning, regional policy, and setting up strategies for employment, the environment, and living conditions. This type of data needs a low speed of information renewal, and limited internal information processing between gathering and dissemination functions.

In the context of European Regional Development Fund (ERDF) innovative actions 2001–2003, the regional administration of Thessaly, Greece, developed a digital infrastructure that included a regional intelligence component. The application was supported by learning networks, which bring together companies, research labs, and business associations to discuss and exchange experience, expertise and best practice. Learning networks were organised in four industry sectors, such as textiles and clothing, food and beverages, construction materials, and furniture. They offered a forum of interaction, information and evaluation of products and technologies selected by company executives. Learning networks were seconded by a Regional Documentation Centre (RDC) that offered information services surveying markets and product innovations. The RDC processed and disseminated information having organised three parallel modules:

- A *technology and market watch* module which searches the web and the technical press daily for new markets and product announcements in the fields of textiles, food, and metallurgy. It continuously updates a database relating to international and financial news, conferences, exhibitions, scientific studies, job opportunities, regulations, new products and technologies, which it offers free of charge to organisations participating in the learning networks.
- A *business benchmarking* module allowing competition analysis. It is an assessment procedure, which compares the performance of a company against a selected group of companies, defining the strengths and weaknesses of the former. For each assessment indicator used, the application provides information about the position of the company within the comparison group and the distance from the best performance. A local network of consultants uses the system and provides the consulting based on the information analysis.
- A *regional innovation performance* module based on annual reports of regional innovation indicators, much like the EU innovation scoreboard. Using the Eurostat database CRONOS, 20 indicators were defined showing the performance of the Region in the new economy. The areas covered are the productive system, human resources, creation of knowledge, and innovation. Comparisons with other regions highlight the strengths and weaknesses of Thessaly and the opportunities offered by the regional environment in which companies and learning networks operate.

RDC is accessible online. In parallel, a newsletter with the latest information update circulates every month. The subject structure is simple and clear, assuming that users will have limited time to spend on the Internet. It was evaluated very positively by its end-users, especially the sections offering information on emerging markets and market opportunities. The weakness is that the system demands continuous updates. A small team works daily to collect, evaluate, and enter data into the databases. The information technology solutions used are rather conventional, with databases and html interface; information search is automated, but most of the work for assessing information is still manual.

Main components of strategic economic intelligence

Sets of information modules make the core components of strategic economic intelligence. These are entities, which gather, process, and disseminate information with specific content and are aimed at defined target groups. The major challenge is to combine knowledge from different sources, to integrate information from a distributed network that collects and elaborates data, and to customise information to the needs of particular users and organisations.

Most important informational modules for building strategic economic intelligence at any level (company, cluster, and territory) are those covering as much as possible key innovation processes, including:

- market and technology watch, which offers updates on technology trends, products, and innovations;
- competition analysis and benchmarking the performance of a selected organisation, company, cluster or region with respect to a defined group of organisations that make the comparison reference;
- foresight about expected changes in regional markets, technologies and socio-economic conditions, and other future trends; and
- research results (R&D, product or service concepts, prototypes, patents) which the user may consult and exploit through technology transfer agreements.

The variety of informational modules composing existing economic intelligence initiatives shows that there is no single solution to the question of structure and content. At least three content layouts clearly emerge: the first focused on the company with the aim of improving its product's characteristics, costs, and placement; the second focusing on industry sectors and clusters to facilitate new product development and market access; and the third targeted at public administrations for improving regional policy and infrastructure planning. These are differences based on the content of the respective information system, but content is not the only important factor. The functionality of any system is determined by a number of decisions related to the selection of information content and modules; the integration and correlation of data from different information modules; the customisation of information for different audiences; the delivery of content on demand with respect to the needs of particular organisations; and the level of automation or the combination of human and artificial intelligence in information retrieval and assessment. Let us take a look at the main information modules.

Market and technology watch

In its simplest form this module appears as collection and dissemination of information about commodities and prices. In more advanced forms it covers product supply and demand, auctions, announcement of new products, new machinery and technology, production reports, and future estimations about prices and production volume. Due to the complexity and extent of information, market watch is better organised on an industry or cluster basis. One of the most sophisticated applications is to be found at www.yarnsandfibers.com, which covers market intelligence on the textile and fibre industry, including price watch, industry news, industry reports, benchmarking indexes, and a trading zone as well.

Market and technology watch requires three operations. First, a continuous and systematic scan of information sources to identify relevant information. Second, entry of this information into a database, analysis, visualisation, and preparation of reports; and third notification of recipients about new information uploaded to the database. Though this looks rather simple, the process is highly sophisticated. Few elements of search and elaboration functions can be automated with the use of crawlers and robots. A team of specialised personnel has to manually perform the necessary tasks and set up intelligent search routines. The level of automation is limited to the spheres of initial data search, data storage, and dissemination via the web.

Competition benchmarking

In different forms, this module is found in most business, cluster, and regional intelligence applications allowing the performance of organisations, companies, and territories to be compared. Benchmarking is a process of identifying, understanding, and adapting outstanding practices and processes found inside and outside an organisation. It is based on the systematic comparison of indicators, which capture the underlying factors of best practice and performance.

Company benchmarking was pioneered by Xerox in 1979 as part of the strategy to cope with international competition in the photocopier market; since then its scope has been enlarged to include multiple business services and processes. The benchmarking process involves comparing one firm's performance on a set of measurable parameters of strategic importance against other firms known to have achieved best performance on those indicators (Kelesidis 2000). There are many ways in which benchmarking can be applied. Competitive benchmarking is performed vis-à-vis competitors and data analysis is done to explore what causes the competitor's superior performance. Internal benchmarking examines differences in performance in organisations that have multiple units or branch plants operating in different regions. Process benchmarking compares discrete process performance and functionality against organisations that are excellent in those processes. Generic benchmarking looks at the way resources and technologies are used at selected companies independent of their industries. The main outcomes of company benchmarking are on the one hand, a definition of strengths and weaknesses of an organisation, and on the other hand, the precise/quantitative definition of improvement margins in management, production, and distribution performance. The results are sensitive to the number of indicators used and the size of the databases supporting comparisons.

On the other side, regional benchmarking has evolved from the analysis of time series on regional development data. Regional statistics on output, employment, and growth offer the basis for describing and modelling changes in regional economies. What is new in regional benchmarking is the simultaneous collection and elaboration of data from many regions and the positioning of a region's performance against other regions. Again the fundamental process is comparison: a set of regional performance indicators is defined, and data allow the changes in these indicators over time and different territories to be traced.

The first attempts to systematically compare innovation and new economy trends on a regional level were made in the US at the end of the 1990s. The *Massachusetts Innovation Economy Index* is probably the oldest exercise, first published in 1997 (John Adams Innovation Institute 2007). It is based on 20–32 indicators (depending on the year) reporting annually on the Massachusetts new economy. It uses statistical data to illustrate how the State performs in the new economy, and compares this performance to selected high-tech states throughout the US. The Index is based on the principle that innovation is a critical factor for development; it focuses on the nine most important industry clusters to better understand how innovation processes influence the growth of these heavily concentrated clusters. All the selected indicators derive from objective and reliable data sources, are statistically measurable on an on-going basis, reflect economic vitality, and measure conditions in which there is an active public interest. Indicators are divided into three interrelated groups:

- Economic impact: Industry Cluster Employment and Wages; Corporate Sales, Publicly Traded Companies; Occupations and Wages; Median Household Income, Manufacturing Exports.
- Innovation process: New Business Incorporations and Business Incubators; Initial Public Offerings (IPOs) and Mergers and Acquisitions; Technology Fast 500 Firms and Inc. 500 Firms; Small Business Innovation Research (SBIR) Awards; Regulatory Approval of Medical Devices and Biotechnology Drugs; Corporate Research and Development Expenditures, Publicly Traded Companies; Patent Grants, Invention Disclosures, and Patent Applications; Technology Licenses and Royalties.
- Innovation potential: Investment Capital; Federal Academic and Health Research and Development Expenditures; Intended College Majors of High School Seniors and High School Dropout Rates; Public Secondary and Higher Education Expenditures and Performance; Educational Attainment and Engineering Degrees Awarded; Population Growth Rate and Migration; Median Price of Single-Family Homes, Home Ownership Rates, and Housing Starts.

Monitoring the State's capacity is crucial for assessing its strength and resilience. At the same time, benchmark comparisons can provide an important context for understanding how Massachusetts is doing with respect to other regions. Massachusetts is compared with the national average or with a composite measure of six competing and leading technology States: California, Colourado, Connecticut, Minnesota, New Jersey, and New York.

The equivalent in Europe is the *EU Innovation Scoreboard*, which was launched in response to the Lisbon Council's expectations of making Europe the most competitive knowledge-based economy by 2010. The first publication in 2001 presented data over the period 1995–2000, while the subsequent annual reports examined various aspects of innovation performance in the EU. The scoreboard provides detailed analysis by country, region, and sector, while as a policy instrument offers new insights on innovation, technology adoption, and development.

The methodological basis of both company and regional benchmarking is more or less the same. The process starts with defining the indicators that we wish to compare; this is then followed by data collection, comparison with data coming from previous periods or other organisations and territories, reporting of main findings, and setting of improvement plans. There are three critical elements in this process. First is the definition of indicators, which should reflect the underlying processes shaping the performance of an organisation or geographic entity. Second, the collection of data over different time scales and territories has to follow the same rules and quality standards. Third, the definition of comparison algorithms should lead to new variables meaningful in the context of models and explanatory schemes.

Foresight

This module codifies data based on foresight exercises. The mainstream form of foresight combines a national and a sectoral dimension. However, foresight has been recently applied at the regional level to better understand ongoing technology trends and improve local decision-making. Regional foresight (RF) has been implemented in Limousin (FR), Lyon (FR), West Midlands (UK), north-east England (UK), Catalonia (SP), the Basque Country (SP), Uusimaa (FN), Central Macedonia (GR), and other EU regions. These initiatives were placed under the innovating regions strategies in the EU regions, which started with RIS projects and continued with ERDF Innovative Actions, and the 'Regions of Knowledge' projects.

RF can be defined as a systematic, participatory process, involving gathering intelligence and building visions for the medium-to-long-term future, and aimed at informing present-day decisions and mobilising joint actions (Miles *et al.* 2002). RF involves thinking about emerging opportunities, challenges, trends and discontinuities; however, the aim is not to produce insights about the future, but to bring together key regional actors and regional sources of knowledge and develop strategic visions and anticipatory intelligence.

Foresight can help regions to break down barriers and to create networks sharing common visions. It can be useful to inform action at any level, from business, to academia and the regional administration. It is only worthwhile when it can be tied to such action. More precisely, regional foresight brings awareness about emerging trends in different areas of regional life, including: *social trends*, with emphasis on human capital, covering issues such as demography, settlement, mobility, identity, citizenship, networks, social capital, education and training, healthcare; *science and technology trends*, with emphasis on emerging technologies, promising R&D, but also market opportunities, social and economic needs; *business dynamics* on major industry clusters, start-ups, economic performance, competitiveness, and exports; and *territorial vision*, in which the region is considered as a whole in a nexus of resources, geopolitics, economy, and development.

Regional foresight involves five essential elements:

- Structured anticipation and projections of long-term social, economic and technological developments and needs.
- Interactive and participatory methods of exploratory debate, analysis and study, involving a wide variety of stakeholders.
- Forging social networks; the emphasis on the networking role varies across Foresight programmes. It is often taken to be equally, if not more, important than more formal products such as reports and lists of action points.
- The formal products of Foresight go beyond the presentation of scenarios, and beyond the presentation of plans. What is crucial is the elaboration of a guiding vision, to which a shared sense of commitment can be attached.
- This shared vision is not a utopia. There has to be explicit recognition and explication of the implications for present day decisions and actions.

(Gavigan *et al.* 2001)

From an 'intelligence' point of view, the contribution of RF is double. First, it goes beyond conventional 'future studies' and brings awareness about long-term trends and challenges into immediate planning and decision-making; thus it links to the preoccupation of cluster and regional intelligence to produce knowledge for immediate action. Second, it recognises that knowledge in the new economy is distributed, and delivers future estimations through networks of experts and participatory consultation. Regional foresight stakeholders are industry associations, universities, businesses, chambers of industry and commerce, technology intermediary organisations, and citizens; the same ones we meet in regional intelligence networks; in this sense RF helps with deepening the distributed and network dimension of regional intelligence.

R&D watch

This module, found in some regional intelligence applications, provides information about the current state of research in public and private organisations. It gathers and disseminates information on R&D and technologies produced by research organisations. The origin of this type of intelligence may be traced back to two global technology digital marketplaces developed by Cordis and Yet2.com.

Cordis Technology Marketplace is a free online service where one can find research and technological development results and search for innovative business opportunities on emerging technologies. It includes exploitable research results stored in the Results Service; a showcase of best results is displayed as technology offers; additional information relates to innovation news, events, useful links, and local support. The information is provided from public and private sector organisations, and from EU and non-EU funded research (regional, national, etc.) as well. Five scientific domains are covered: Biology and Medicine, Energy, Environment, Information Technologies and Telecommunications, and Industrial Technologies. Technology offers are classified into three areas, according to offer marketability and closeness to market exploitation:

- Business offers, which are close to market exploitation and for which a prototype has already been developed.
- Science offers, which are at the research and development stage; it is highly scientific in nature and has exploitable potential for a very selective/specialised market.
- Society offers, which are involved with concerns/issues that affect society at large.

All results included in the market place are awaiting further exploitation, such as production and/or marketing agreements, further development or funding. The database is updated whenever new results become available (usually on a weekly basis). Entries are comprehensive, providing information about the research result, the contributing organisation, and the type of collaboration sought, prototype availability, commercial potential, contact point information, and other details.

Yet2.com is the first global forum for buying and selling technology on the Internet. It was founded in 1999. A self portrait highlights Yet2.com as a virtual technology marketplace, offering companies and individuals an unprecedented opportunity to conveniently and privately purchase, sell, license and research some of the world's most valuable intellectual assets. Spanning all industries and areas of research and development, Yet2.com is a community where technology officers, scientists and researchers can unearth cutting-edge discoveries as well as new applications for tried and tested technologies. Yet2.com helps companies extract value from undervalued or unused technologies by streamlining the traditionally lengthy and ineffective process of technology transfer. Many of the world's premier research and development companies currently provide proprietary technologies on an exclusive basis to Yet2.com, creating a robust marketplace where the world's most coveted inventions are listed, sold and, ultimately, applied.

At the regional level, few R&D intelligence applications have been developed, among which probably the most comprehensive is 'Madri+d' in Spain. *Madri+d* is the regional information and technological promotion network for public research centres and private non-profit entities linked to the technological innovation of the region. The network is composed of 35 organisations with 14,000 researchers and it is coordinated by the Comunidad de Madrid. Madri+d focuses on managing and disseminating the intellectual capital of regional institutions and companies by intensively using information technologies and the Internet. It also works on defining common strategies and methodologies in the exploitation of research results, providing high added-value services to researchers and companies, and the motivation for the creation of new technology-based firms. Overall Madri+d manages regional scientific and technological knowledge, adding value to territorial competitiveness, and allowing the public to obtain part of Madrid's science and technology issues (Madri+d 2007).

The advantage of regional technology marketplaces with respect to global applications is in the post-information stages of technology transfer and absorption. Cooperation between technology providers and users is more easily developed on a regional rather than international scale. Regional technology exploitation networks seem more effective. The reasons are well justified in the literature analysing the geographic scales of technology cooperation and transfer, and the problems produced by the geographical, cultural and linguistic distances between technology providers and users (Gentler 1996).

If technology cooperation is more effective at a regional level, regional R&D databases suffer from limited diversity. However, the scope of regional technology marketplaces is to counterbalance the lack of internal R&D departments in most companies. Providing information about research capabilities and results that are available next door is the first move towards substituting the internal R&D departments of companies by external public and private research.

Integration of distributed intelligence

We now have a clear picture of two things: first that economic intelligence gathers data coming from different information modules with specific focus and perspective; and second that economic intelligence changes with respect to the recipient, be it a company, an industry cluster, or a region. However, this system presupposes integration of the data provided by the different information modules. The greater the integration is, the more advanced intelligence we get.

Clusters and regions offer the background for integration. In fact, territories have always worked as 'integrators' bridging individual skills and developing functions for cooperation and coordination. Looking at a number of regions offering innovation-friendly environments, we come across several information systems facilitating decision-making and innovation, including regional foresight, company and regional benchmarking, R&D databases and interfaces matching technology demand and supply, online technology and market watch, as well as databases with skills and competences information. However, these systems are usually disconnected; they are not integrated; each one is addressed primarily to a different group of users and provides information in a specific field of interest. For instance, an application that produces benchmarking reports on

Intelligence and integration

According to Minsky (1988) integration is a key process in human and artificial intelligence. In the 'Society of Mind' he argues that intelligence is the result of the interaction of a vast number of distinct and individually simple, but intricately connected, processes known as agents that are themselves mindless. In this sense intelligence emerges from non-intelligence. He calls this structure of rudimentary non-intelligent agents, which combine and link together to form broader, higher levels of complexity the 'Society of Mind'. Each agent by itself can only do some simple thing that needs no intelligence or thought at all. Yet when these agents are organised into societies – in certain very special ways – this leads to intelligence. Agents are ordered in agencies which are structured sets and can carry out functions different from the parts comprising them. Agencies can use other agencies without comprehending how the latter function. This is the most normal relationship between agencies. Intelligence then is a network and hierarchical tree; it does not arise from certain isolated, specialised processing centres but from organising a large number of non-intelligent particles.

In newer views about intelligence, the concept of integration became more important than in Minsky's hierarchical ordering of agents. Associative structures and distributed information networks may generate exceptionally complex functions which allow increased problem-solving capabilities. Connectionist models for information processing have a number of advantages for engineering and other types of AI applications. Comparing traditional AI and connectionist approaches Kennedy, Eberhart and Shi (2001) outlined some major differences in the way objects are represented, stored, and retrieved. In the case of traditional AI, an object is represented by a static symbol; storage is storage of symbols; and learning is the reconstruction of a symbol. In connectionist AI, an object is represented as a pattern of activations across the network of processors; storage is the matrix of weights between pairs of processing elements; and learning is the adjustment of weights connecting the processing element. Both symbolic and connectionist models, they argue, explain some aspects of human intelligence and solve certain engineering problems, and the matter is applying the right method to the particular instance. The symbol-processing paradigm has been pretty good for resolving well-defined problems like chess playing, but it is totally impractical for noisy, ambiguous, non-linear data or poorly defined problems. The connectionist model avoids strict logical formalism allowing learning from use, trial-and-error, and a more humanising use of machines.

158 Building blocks of intelligent cities

business performance in a given industry does not offer the opportunity to check what the critical technologies in the industry are nor what the market trends and market niches available are. Informational integration, joint research capability, interoperability across applications and types of information, are qualities which are missing in most existing systems of regional intelligence. However, we cannot speak properly of cluster or regional intelligence until this integration takes place.

Integration is crucial for any structure managing information, cognition, and knowledge. In the field of economic intelligence, Figure 6.1 shows how we might bridge and integrate distributed informational modules located in a region or a cluster. A core is created with the connection of public domain databases and content, which are operated by different regional actors and providers. The regional/cluster dimension assures the compatibility of modules and the organisational base and trust that are necessary for this system. Integration is about providing common entrance gates, search functions, definition of inter-database descriptors, compatible content categories, meta-data, etc. The periphery is made up of individual databases belonging to companies, which perform their own data mining, scorecard, modelling and reporting functions with respect to internal data and external information from core public domains. Integration is two-fold: on the one hand there is integration between the core modules of

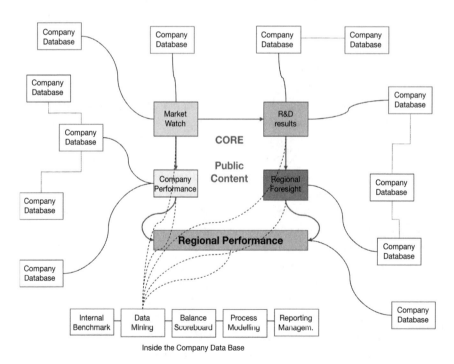

Figure 6.1 Integration of informational modules

market watch, company performance, R&D results, regional foresight, regional statistics, etc., and on the other it takes place between public content and company data.

Various strategies may be deployed to make this integration happen. A precondition is to have agreement between the stakeholder organisations about linking their information resources and adopting common collection, information processing, and dissemination procedures. Strategies differ with respect to the type of agreements and the technologies applied.

Ex-ante integration is centralised integration and presupposes early stage coordination between the stakeholders of information modules. The agreement starts from the design of the core informational modules in order to allow interoperability and common standards in data entry, data communication and exchange, and search functions. Ideally it would be a unique regional/cluster database with different sections according to the modules selected. Organisations participating in the development of the system specialise in separate informational modules, while cooperating in managing the global system. Specifications are communicated to companies on how to harmonise their internal databases with the core. A central coordination agency is needed to control integrated design and resolve problems created from differentiation, maturing, and improvement of core modules.

Meta-search integration is lower level integration, less centralised, but more open and expandable. Organisations develop the core information modules separately, but design agents and integration servers to work on their system. Exploiting the increasing amount of diverse web-accessible data is recognised as an important problem. In this context, data modelling and related knowledge processing need a comprehensive representation and modelling of metadata, potential use of knowledge representation for correlating metadata from heterogeneous media, and use of ontologies to deal with terminological differences between terms in the information requests and those in the metadata and data (Shah and Sheth 2000). Integration servers may also contribute, enabling effective analysis covering many applications and databases, integrating a large number of tables into logical models, allowing the user to search simultaneously in structured information as well as unstructured information.

Creative integration is a step forward, but it is less structured. It is based on a number of informational modules, but also includes an active information processing team that creates new content by combining distributed data. The system becomes proactive with the publication of periodical reports, newsletters, personalised information bulletins, while new descriptors and indicators become available from integrating distributed data. Creative integration uses both advanced information technology and human intelligence. From the collection and elaboration of data, it goes on to show good practice on how to use information, indicates links and associations in the interpretation of data, and develops creative thinking on how to proceed from information and learning to innovation.

Meta-foresight: an experiment in strategic economic intelligence

An attempt to develop strategic economic intelligence was made as part of the Meta-foresight project. The acronym denotes both the use and further advancement of knowledge generated during regional foresight exercises. The objective was to create an integrated strategic information system for market and technology watch, based on cooperation among university and research institutions, private companies, sectoral associations, and public authorities located in a region or sustaining a cluster. Meta-foresight was part of the first generation of 'Regions of Knowledge Pilot Actions' introduced by the European Parliament in 2003, which later became part of FP6 and FP7. RegKnow aims to support experimental actions at the regional level, to develop 'regions of knowledge' with a strong capacity for technological development, cooperation between universities, research and innovation at a regional level, and to stimulate the integration of regions in Europe (Cordis 2003).

Meta-foresight brought together skills and information management expertise from five R&D organisations working in different regional contexts: the URENIO Research Unit (Aristotle University of Thessaloniki, Central Macedonia, Greece) as coordinator, FUNDECYT (Foundation for the Development of Science and Technology in Extremadura, Spain), University of Wales, Cardiff (Wales, UK), INFYDE (Informacion y Desarrollo S.L., Basque Country, Spain), and the Institut Jules-Destrée (Wallonia, Belgium).

The main concern and core concept of Meta-foresight was to integrate information from five information fields and offer intelligence at business and cluster (sector) levels. The five fields are:

1 Regional foresight, which allows systematic, participatory, future intelligence gathering and medium-to-long-term vision-building and mobilisation of joint actions.
2 Benchmarking that fosters learning from others by comparing practices and performances.
3 Market watch, which provides information on product supply and demand, new products, prices, emerging markets, and channels of distribution.
4 R&D watch, which focuses on technologies emerging from regional and global R&D players, and identifies patents and other IPR enabling the acquisition of promising technologies.
5 Regional technological competences and skills, which allow available expertise in the region to be identified to support innovation and technological solutions.

Meta-foresight has developed a concept and created the respective information system of collective intelligence. Cooperative collection of data is the cornerstone underpinning the solution and requires the active participation and collaboration of many regional organisations and users. For this purpose a network of actors has to be established within each region/cluster where it is being implemented in order to continuously feed the application with data and share information and knowledge.

We should stress at this point that information intelligence, learning, anticipating the future, defining ways to act in fuzzy situations, are at the core of what we usually call individual intelligence. The term describes our capacity to learn, to solve problems, and plan for the future. However, the same qualities can be found in human collectives, which also exhibit capacities to learn, to solve problems, and adapt to external constraints. Collective intelligence characterises social entities, like groups of people, networks, clusters, organisations, communities, cities and regions, acting intelligently as a whole to address problems and questions beyond the reach of individual capabilities and means. Collective intelligence is collaborative intelligence; it means working with other people, sharing information and capabilities, establishing common goals, learning from each other. Collective or collaborative intelligence materialises as a network of cooperating entities, including individuals and organisations, operating along defined rules that assure superior capability and achievement in the field of information and knowledge compared to the individual capabilities of the network members. Thus collective intelligence emerges from the combination of individual actions by following predetermined rules. It may comprise a small or large number of cooperating entities, depending on the capacity of its constituting rules to incorporate a smaller or larger number of entities. Clearly it is a systemic phenomenon, in which much of its capabilities do not reside in its individual parts but in the system's characteristics, working patterns, rules and institutions. 'Collective intelligence of the whole is greater than the collected intelligence of its parts because wholeness adds synergy – patterns of relationships and interactivity – to the mere sum of parts' (Kennedy and Eberhart 2001).

Meta-foresight adopted the above collaborative/collective intelligence viewpoint and tried to devise an operational information system for strategic intelligence at the level of company and cluster/sector. Collaboration and integration, as core concepts of Meta-foresight, had two complementary sides. On the one hand the system is based on the cooperation of information providers, referring to a combination of information and knowledge from organisations active in the above five fields of intelligence (foresight operators, benchmarking agencies, markets, R&D, and competence observatories). On the other hand, it is based on participation of users in the assessment and flow of information; a feedback from users leading to better integration of information between providers and recipients.

The Meta-foresight platform

Information integration from the five thematic fields mentioned above is guided and facilitated by a software platform and toolbox. The core methodology for integration is benchmarking. Different areas of intelligence (audits, regional statistics, market watch, etc.) are compared and conclusions are drawn.

162 Building blocks of intelligent cities

Foresight also introduces comparisons between actual facts gathered from audits and data from market watch and future trends. The platform connects the supply of information to the demand for, and the management of, information.

On the supply side, the platform brings together a network of organisations located in a region or cluster that have developed and are operating applications in different fields of information. Authors from these organisations provide information to the system and comprise the nodes of the network. This open architecture allows for many organisations, experts, and interested individuals to take part as active information collectors and writers.

On the demand side, the platform connects to a network of information recipients. These are individuals in different target groups, companies, intermediary organisations, consultants, managers in clusters and regional authorities, each of which has specific information interests and needs. Data from the users are combined with data from the providers, producing more accurate and customised analyses and reports.

Between the information supply and demand stands the Meta-foresight back office, which operates the platform. The back office sets the scene of this collective intelligence attempt: assigns 'authors' and 'recipients'; defines the rules for information collection, analysis, and dissemination; and customises the platform's tools to the local information framework.

The overall structure of the platform and the network that is sustaining it is presented in Figure 6.2 showing the constituent elements and operating lines.

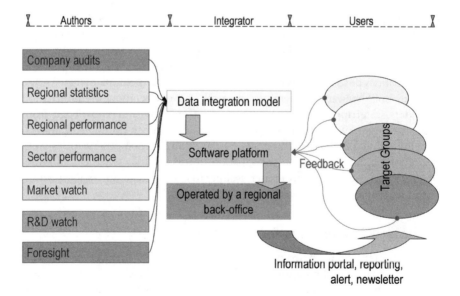

Figure 6.2 Meta-foresight structure

Strategic economic intelligence: global watch   163

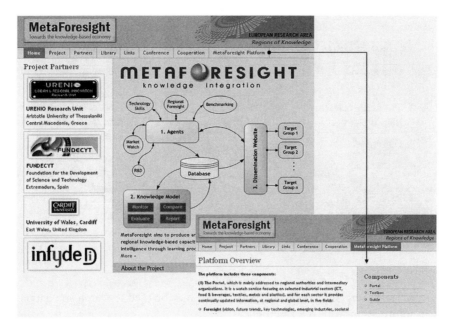

Figure 6.3 Meta-foresight homepage
Source: www.urenio.org/metaforesight

The platform is web-based and both authors and users can contact it online. This solution easily allows the application to be shared between the institutions that have agreed to cooperate and combine their internal information repositories. On the other hand, it facilitates dissemination of information to recipients, while offering online information on all existing information packages. Access to the platform is available via the Internet at the address www.urenio.org/metaforesight.

Nine distinctive tools for information processing are available and operate on the platform. They can be found in the three areas of the platform, corresponding to information collection, analysis, and dissemination. These are usual functions of business intelligence tools and Meta-foresight has adopted the same structure.

Data collection is the process of defining information sources, 'visiting' them, and extracting information. Any member of the information provider network can use the data collection template on the platform, fill it out, and submit it to the system. Four distinct tools (T) facilitate data collection:

- T1: A *web-based list of information sources*, regional, national or global corresponding to the Meta-foresight information modules. Apart from these sources, external business intelligence tools, equipped with dedicated agents

164 *Building blocks of intelligent cities*

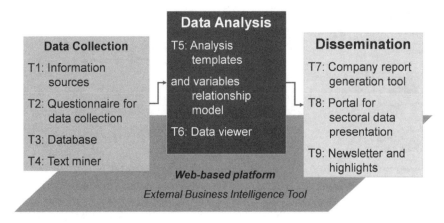

Figure 6.4 Meta-foresight toolbox

for searching different types of databases, may be used to search and gather information from the Internet.

- T2: A *data collection questionnaire*. This tool is useful for the Meta-foresight business intelligence application, the Confidential Company Report (CCR). The questionnaire is divided into eight thematic fields: (1) financial performance; (2) strategy and management; (3) products; (4) markets and competition; (5) research and innovation; (6) production processes; (7) supply chain and networks; (8) quality and standards. Preparation of the CCR starts with a company audit, which highlights the main issues affecting its performance. Information from the company audit is combined with data concerning the region and the sector which the company belongs to, the markets in which it operates, related technologies, and promising research. The purpose of the tool is to exactly define the targeted information, and the quantitative and qualitative variables that are used in order to collect data.
- T3: A *database to store information*, both internal information concerning the company, and external about markets and technologies. The structure of the database corresponds to the structure of the questionnaire. Administration of the database is flexible and changes may be introduced during the customisation of the platform for different regional contexts.
- T4: A *text-miner* to facilitate the transfer of any piece of useful information located in the sources to the database. The text-miner is a tool that allows one to mark useful pieces of information on the Internet, mainly comments and qualitative comments, and then transfer them directly to the database, creating a separate record for each entry, while preserving information about the data source. For each variable more than one record can be made in the database.

Data analysis follows on from data collection. This is primarily the process of going deep into data, integrating data, combining available information in meaningful ways. Data analysis is the core of Meta-foresight intelligence. It is based on human skills used mainly to compare and comment on data. Two dedicated tools assist and facilitate data analysis:

- T5: A series of *analysis templates* (eight in total) allow for the integration of content included in the database. The templates correspond exactly to the themes of the questionnaire and the sections of the CCR: (1) financial performance; (2) strategy and management; (3) products; (4) markets and competition; (5) research and innovation; (6) production processes; (7) supply chain and networks; (8) quality and standards. Within each theme, a step-by-step procedure guides the analysis from the company to the region, the sector, the market, R&D, foresight, and competences. Precise questions, benchmarking results, and free spaces to comment on quantitative and qualitative data allow information to be combined and integrated. A default form is provided, while the administration module allows the content of templates to be modified. Each template follows a step-by-step procedure, with predefined questions. For each question it defines which variables from the database will be correlated, and which subjects should be addressed.
- T6: A *data-viewer*. This tool provides access to the content included in the database. The purpose here is to allow a partial and step-by-step consideration of the information collected, relating it to the question to be answered and the internal procedure of analysis.

Dissemination is the third part of the platform. It supports three applications, the Confidential Company Report and the Portal on sectoral information. Dissemination is confidential in the case of the CCR, and open to the public in the case of the Portal.

- T7: The *Confidential Company Report* is based on the work done during analysis, and the step-by-step elaboration and completion of templates. The report follows the structure of the templates, but also includes benchmarking material, diagrams, and comments made by the Meta-foresight team.
- T8: The *Portal of Sectoral Information* offers information and news for up to five industry sectors/clusters per region. For each sector the Portal provides continually updated information, in the form of stories, on (1) foresight (vision, key technologies, emerging industries, societal trends); (2) R&D and innovation (current research, patent information, new processes/technologies, regulations and standards); (3) market trends (news, prices, trends, market analysis/reports); (4) benchmarking (regional index, industry indicators, best practices, competition practices); and (5) regional competences (centres, labs, experts, suppliers).

- T9: The *Newsletter* is created automatically from the stories on the Portal, usually the most recent ones in the sector. An online application facilitates the creation of electronic alerts and newsletters. The purpose is to make information on the Portal known to a wider public. The newsletters are sent by e-mail to the list of recipients and users.

These nine tools are at the disposal of the Meta-foresight back office team that coordinates and manages information processes. The platform is aimed at two target audiences:

- cluster managers and intermediary organisations, wishing to offer **sectoral/ cluster intelligence** reports and analyses; and
- companies and consultants wishing to apply **company intelligence** and offer market and technology updates, benchmarking reports, and other types of intelligence taking into account emerging trends and available regional skills and competences.

Sectoral/cluster intelligence

Sectoral intelligence entails systematic monitoring and dissemination of information concerning an industry sector or cluster. In the case of Meta-foresight it is provided through the operation of a public Portal organised per industry sector and thematic module.

The Portal homepage presents the most recent entries for each sector, dealing with technologies, markets, prices of raw materials, R&D, patents, etc. Lists of regional companies per sector as well as subcontractors are also presented. The Portal stands on the cross-roads of two networks of 'authors' and 'users'. Between are the administrator and the Meta-foresight back-office.

The **authors** are physical persons from the network of cooperating organisations with competences in the fields of foresight, innovation, market watch, benchmarking, as well as from the main regional competence centres. This network of authors is the core of sectoral intelligence. They feed the application with relevant information. The author, after login, selects the industry sector in which he/she intends to write a 'story' and goes on to provide the title, the summary, the full story, and the reference text. A set of meta-data also characterises each story. A series of buttons is available for formatting the text. Then after selecting the category and sub-categories to which the story belongs, he/she saves the entry, and logs out.

The **users** are regional authorities and institutions, technology intermediary organisations, cluster managers, companies, and physical persons, who can scroll through the stories available or search for a specific topic in the Portal database using the search engine provided. The users' feedback is vital in guiding the selection of stories and highlights on the Portal. Users receive information from the stories written by the authors. The Portal offers a newsletter facility, collecting information from the 'stories' and communicating them via e-mail.

The **administrator** is part of a team of experts responsible for managing the whole application, coordinating the regional technology cooperation network to collect 'stories', validating them, and maintaining the Portal. The administrator can assign authors and provide login passwords, edit or delete authors' stories, add new categories or remove a category, write or edit articles, add or remove members, create and send the newsletter. Overall, administration of the Portal is extremely friendly to make data entry as simple as possible, while only demanding minimum knowledge of programming.

The Portal allows a collective watch function over an industry sector to be performed. It is the outcome of a network of actors that cooperate to reveal what really happens in that sector. Stories are about products and processes, market changes, emerging markets, price anticipation and future estimations, informed opinions about future technologies, and highlights from achievements in science and technology in regional competence centres and clusters. The complexity and breadth of information makes the cooperative input necessary and a very cornerstone of the application.

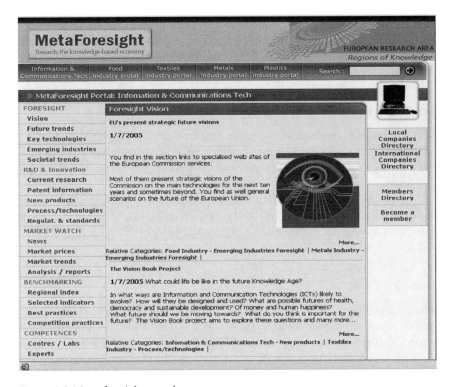

Figure 6.5 Meta-foresight portal

Source: www.vrc.gr/metaforesight/portal/par_kain.aspx?sect=goun&catid=82&page=0

The collective data input enables information produced by regional foresight, R&D, benchmarking data, market watch, product and process innovation to be gathered in order to provide information about the current situation, trends, and opportunities in the region. The network of regional actors is more important than the software application itself. The Portal is just the medium that stores and disseminates the results of the regional collaborative effort to gather information and disseminate it.

Another side of this collective watch of products and technologies is the evaluation provided by users. Information on the Portal is aimed at a large number of recipients involved in the activities of the sector. There is therefore a user evaluation template, in the form of a comment, for all entries. After having read the entry the user can assess it and send the evaluation back to the system. Processing of assessments enables a benchmark to be assigned to each entry, allowing for its significance with respect to this benchmark to be reconsidered, and for the network of authors to be informed accordingly. A neural network may be used to estimate weights in the variables used to construct a composite benchmark from a series of evaluation criteria included in the template. This solution of assessment procedures, which combines benchmarking and neural network techniques, was introduced by the work of Choy et al. (2003) who applied the methodology for assessing suppliers of large corporations. The method makes use of connectionist approaches of artificial intelligence applied in cases of network structures. The large number of information recipients and their feedback makes it feasible to rate the information they receive and introduce this rating into the functioning of the system.

Strategic business intelligence

Strategic business intelligence is a complementary application supported by the Meta-foresight platform. The focus of interest in this case is the company (in the manufacturing or services sector) and the information about its main fields of activity.

Most of the tools available on the Meta-foresight platform are used to provide company intelligence on markets, technologies, and future trends. Seven tools combined lead to a Confidential Company Report (T1, T2, T3, T4, T5, T6, and T7).

Starting a CCR

The creation of a new company file is the starting point in generating a CCR. At this point, the newly created report is just blank. What is important – and what is secured at this stage – is confidentiality. The CCR is for the eyes of the author only. No other user of the Meta-foresight platform has access to the particular company report and its successive versions up to the final one. Once a new CCR has been created, the author of the CCR starts with data collection and entry.

Data collection

The questionnaire (T2) drives data collection. It is focused on the principal areas of company activity, highlighting the main issues that a company wants to know about in order to improve its technological and innovation capabilities. The default template covers eight areas: (1) financial performance; (2) strategy and management; (3) products; (4) markets and competition; (5) research and innovation; (6) production processes; (7) supply chain and networks; and (8) quality and standards.

The questionnaire is extensive, featuring 373 questions in the above eight thematic areas. Data are quantitative and qualitative, and in many cases take the form of comments. However, there is no need to collect all these data to generate a CCR; even with a smaller number of questions, the company gets a satisfactory report. The Meta-Foresight team has to decide on the extent of the questionnaire and customise it accordingly, excluding secondary or unavailable information.

The sources from which data are collected are both internal to the company and external as well. At the beginning the CCR examines company activity as codified in a company audit. Then, the information from the company audit is combined with data concerning the region and the sector to which the company belongs, actual and potential markets for the company's products, related technologies and promising research results, relevant trends from foresight exercises, and competence centres in the region.

Data for the CCR cover all major fields of company activity, but above all cover the issues related to innovation, technologies, and processes. All information obtained during data collection is stored in a database (T3).

Table 6.2 Data collection: questions per subject and source

Questions	Source							
	Company	Region	Sector	Market	R&D	Foresight	Competences	Total
Finance	12	9	12	6		8		47
Strategy	21	5	21	3			1	51
Products	8	11	6	2	3	3	4	37
Markets	15	3	15	6		2	3	44
R&D, innovation	15	14	10	3	3	5	4	54
Production	27	8	18	4	3	2	4	66
Supply chain	16	3	16	2	3	2	4	46
Quality, standards	10		10	2		2	4	28
	124	53	108	28	12	24	24	373

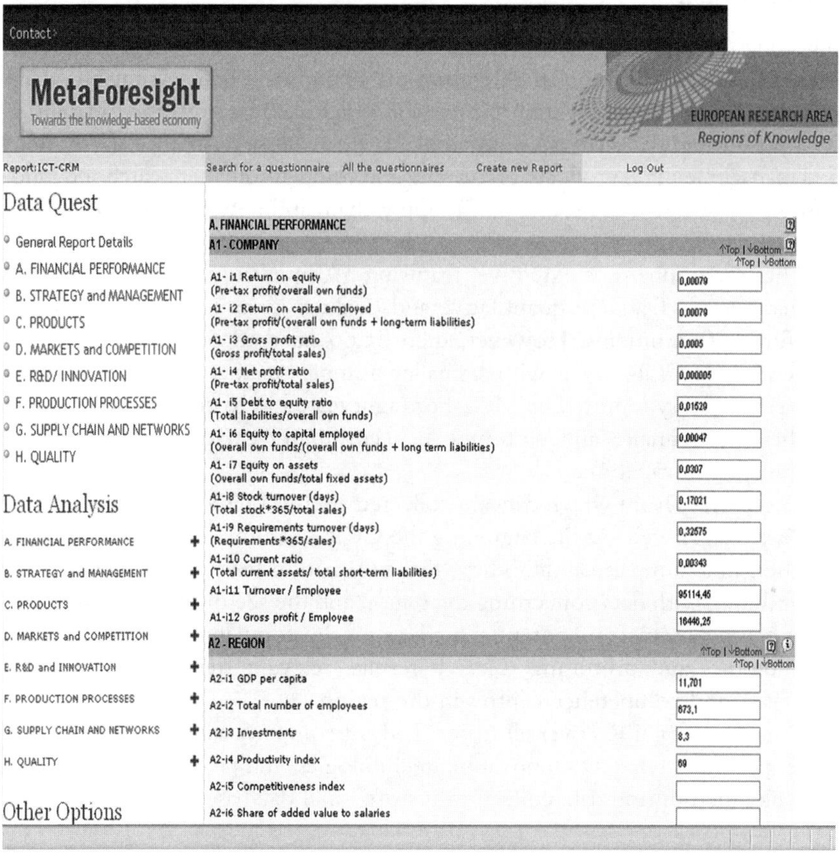

Figure 6.6 Data entry and analysis tool
Source: www.vrc.gr/metaforesight/Default.asp?LangID=2&AppID=ToolBox

The text-miner tool (T3) facilitates data collection on the Internet. By typing an Internet address on the miner, the respective web page appears and the operator has the ability to mark pieces of information and transfer them directly to the CCR database. The miner creates a separate record for each entry, while preserving information about the data source.

Data analysis

Data analysis is a process simultaneous to the preparation and writing of the CCR. As with data collection, analysis is organised in eight sections, which are the same as the sections for data collection. A template guides data analysis in each of the eight sections. In sum, eight templates guide data analysis. The templates are structured around themes and questions to be filled out using free text or numeric data. Each of the eight themes is examined from seven different perspectives: the company, the region, the sector, the market, R&D, foresight, and available skills and competences in the region (Figure 6.7). An evolutionary workflow, in each template, drives the analysis through successive stages examining the company, the company within the region, within the sector, the market, relevant R&D, foresight, and regional competences.

When it comes to benchmarking issues, the template automatically displays the available data and creates the corresponding comparison tables. Additionally, a data viewer (T6) is available, which enables the reporter to consult the database at any point during preparation of the CCR.

Reporting

With the conclusion of data analysis most of the CCR is already prepared. The CCR is produced automatically and stored in the database. As mentioned, the CCR created is for the eyes of the consultant who wrote it and the company to which it is addressed only. A print preview function enables one to view all sections together. The author may select the benchmarking comparison group from the companies included in the database. A wizard drives the selection using the country, the region, and the sector (given by NACE). Save or Print functionality marks the end of the report.

A full account of the Meta-foresight platform and toolbox is given at URENIO (2007a).

	Perspective	Information source	Type of data	Focus
(1) Financial performance (2) Strategy and management (3) Products (4) Markets and competition (5) Research and innovation (6) Production processes (7) Supply chain and networks (8) Quality and standards	COMPANY	Company audit	Quantitative and Qualitative	Company practices
	REGION	Regional statistics, regional benchmarking, and regional development programmes	Quantitative and Qualitative	The company within the region
	SECTOR	Sectoral benchmarking	Quantitative	The company within the sector
	MARKET	Market watch	Quantitative and Qualitative	Most important facts related to the sector and the products of the company
	R&D	R&D watch	Qualitative	Most important R&D related to the sector and the company practice
	FORESIGHT	Foresight exercises	Quantitative and Qualitative	Future trends related to the region and the sector
	COMPETENCES	Regional competence centres	Qualitative	Regionally available skills, and competences

Figure 6.7 Data analysis: eight themes under seven perspectives

Strategic economic intelligence: the search for the 'next big thing'

Meta-foresight is an example of strategic economic intelligence at business and cluster level, integrating information from regional, sectoral, and company perspectives. Information from different sources (company, region, sector, market, R&R, etc.) is combined together to give a holistic view of a subject (strategy, innovation, quality, etc.). The system adds intelligence because of the combination and integration of information. Its network philosophy is important for gathering information and getting feedback from users; linking supply and demand aspects of the system; keeping the system alive and close to needs; keeping it down to earth. The digital space primarily equates to the tools for guiding and storing economic intelligence; it is what facilitates smooth operation of the system, but also what helps it reach a wider audience; it is also a tool for analysing how users treat information, what their entry points are, what information they consult, the trajectories and exits they follow. Understanding better ensures added value and promotes improvement.

Economic intelligence has become crucial for innovative companies and territories. The *BusinessWeek Magazine* (2003) dossier 'The Future of Technology' illustrated this trend well. Investigating the prospects for the next round of technologies after the collapse of the dot-com boom, it looked at Silicon Valley, which faced an extremely severe recession, cutting back about 20 per cent of its workforce and 22 per cent of employment in the software industry:

> Although it has never before been taken down so hard or for so long, the local economy has always experienced wild booms and busts. In the mid-1980s, when the PC industry consolidated around a handful of companies, the area lost nearly 10 per cent of its jobs. It didn't dip below 5 per cent until the month after Netscape went public. The Web breathed new life into the Valley. What followed was a historic run. By December 2002, more than 140,000 jobs had been created in the San Jose area, according to the Labour Dept. Then the bottom fell out. By April, 2003, eight years' of job creation had been wiped out.
>
> (BusinessWeek 2003, pp. 42–3)

The focus of the dossier was to elucidate 'what the next big thing' would be and whether a new round of technological innovations would lead this region to thrive again. However, what the stories established was that no-one could convincingly tell what the 'next big thing' would be; or furthermore, whether the Valley would be part of it.

This kind of situation is typical of regions following a knowledge-intensive development trajectory. The advantages offered by innovation breakthroughs very soon evaporate because of duplication and relocation of the same factors that initially contributed to breakthroughs. The region has to go on searching for a new round of innovation and competitive advantage. Ceaseless ups and downs are compressed in cycles of technology investment and disinvestment. Surfing on waves of innovation, an intelligent region has to assure that when something new happens companies and technology organisations will be able to respond to it.

Strategic economic intelligence is part of the new arsenal of innovation-led regional development. But, organisations and territories do not need to reinvent the wheel. The first move is just to keep up-to-date. Integrating information, knowledge and competences distributed among organisations and individuals over a territory ends up opening minds to emerging social behaviours and trends vis-à-vis technological innovations. Technologies live in laboratories and research institutes waiting for social demand to make the 'wave function collapse' and crystallise a new round of innovation. The transition to a global economy seems to be more disruptive in its reach, scope and scale than prior waves of innovation and production restructuring.

7 Technology transfer and acquisition
Virtual spaces valorising academic research

Getting state-of-the-art technology

Among the main tasks of strategic economic intelligence is the collection and dissemination of information about the state-of-the-art in technologies, products, and business processes. For instance, which technologies are important to produce better quality and healthy food using biomaterials, bio-diagnostics, and intelligent packaging? Which steps are important to turn to advanced renewable energy sources – anticipating rising oil prices – including photovoltaic, hydrogen and fuel cells, wind turbines, biomass and biogas? Which technologies offer clean, cost-effective, and low-capital intensive production in the steel and construction materials industries? Which steps are necessary to shift textiles towards synthetic fibres and wearable electronics? Which technologies can strengthen intelligent manufacturing, reduce time-to-market, and help globalise supply chains?

The next challenge, which we will discuss here, is getting what strategic intelligence has brought to the surface: acquiring, absorbing, and using state-of-the-art technologies through technology transfer, licensing, and cooperation networks.

Technology transfer is indispensable in any territory adopting innovation-led and intelligent city strategies. Among the objectives of such strategies would be to assess the technologies in place and take the necessary steps for getting state-of-the-art technology appropriate to existing clusters and sectors. It is important to stress that most productive activities of a region rely on technology transfer, as technologies come mainly through technology transfer channels, purchases of machinery, imitation, licensing, and cooperation with more advanced producers, universities, etc. Innovation is the exception rather than the rule. It is not feasible for a territory to be innovative and a leader in every sector. Silicon Valley is good in semiconductors and Internet services, but falls behind in the automotive, aerospace, and wireless communication sectors, in which Stuttgart, Toulouse, and Helsinki have the lead.

State-of-the-art technology comes embodied in machinery, capital goods, patents, blueprints, standards, prototypes, and industrial designs. It also comes in the form of tacit knowledge embodied in the person's mind and is 'transferred' only via close interaction, learning by cooperating, and mainly through the mobility of personnel. Acquiring state-of-the-art technology is an important threshold to innovation. It corresponds to innovation-to-the-company, adopting innovations that have been tested and proven to be successful, as opposed to innovation-to-the-market, which goes beyond the state of the art. However, innovation does not necessarily pass through the state-of-the-art. Shortcuts are not only feasible, but also usual.

Thus technology transfer and technology absorption are complementary, but different to innovation. It is about using available technologies and producing products already on the market. On the contrary innovation, in the full meaning of the term, is making something new that never existed before. The line between them, however, is not that clear and easy to define. As opposing manifestations of 'imitation' vs. 'creation' the distinction is obvious, with technology transfer falling on the side of imitation and innovation on the side of creation. However, there is also a large grey area between them, where the two practices overlap, as in the case of simultaneous imitation and adaptation; creation based on imitation; imitation transfer to a different sector; part imitation and part creation.

The grey area also comprises uses of the term 'technology transfer' as equivalent to 'innovation'. We learn in Wikipedia, for instance, that 'Technology transfer is the process of developing practical applications from the results of scientific research' (Wikipedia – Technology transfer 2007). But, this is exactly the meaning of innovation: turning scientific knowledge into practical products. However, in the case of academic research, this type of innovation is technology transfer too from the university to organisations that buy and use academic intellectual property.

The landscape of technology transfer

How is technology transfer and absorption achieved? Can an organisation just go out and buy what technology it considers important and wishes to apply?

The fourth annual conference of the Association of European Science and Technology Transfer Professionals (Copenhagen, 22 and 23 May 2003) has provided a good picture of the practices associated with technology transfer. The theme of the conference was 'Best Practices in Transfer of Science and Technology' and the sessions were organised along the classical distinction between technology licensing and technology dissemination. This distinction concerns the institutional aspects of technology transfer and, together with arm's length exchange of technology, shapes the entire landscape of technology transfer. The latter looks something like the following tree (Figure 7.1).

There are important differences between technology transfer routes. Among the three main routes, inter-firm cooperation is under the auspices of the market; it is mainly an arm's length relationship. On the contrary, university–industry cooperation is an institutional relationship. Licensing covers both others. It is

176 Building blocks of intelligent cities

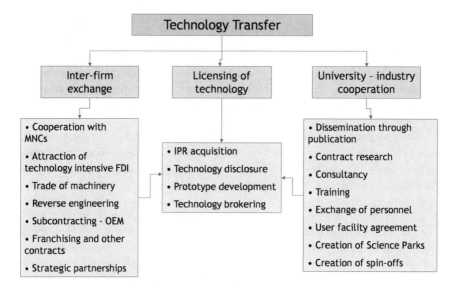

Figure 7.1 The landscape of technology transfer

necessary in inter-firm and university–industry transfers. On any route of the technology transfer tree two issues co-exist. On the one hand, there is technical concern with and assessment of the technology to be transferred, including its technical adaptability with regard to the envisaged utilisation, customisation before it can be applied by the technology user, and potential use in other industrial sectors than the originating one. On the other hand, there is the contractual aspect of the transfer, which mainly concerns IPR, licensing, knowledge propriety rights or arm's length relationships. Non-contractual forms, such as dissemination through publication, migration of personnel, and informal networks, may in certain cases transfer extremely valuable technologies.

From the point of view of technologies, the itinerary followed has an impact on the technology acquired. For instance, the newest and most valuable technology is internalised, taking place within multinational companies (MNCs) or joint ventures with MNCs, while other less crucial technology is licensed.

From the point of view of organisations, MNCs are the major source of proprietary technologies and dominate technology flows in all forms. However, MNC export activity is taking new forms within global production networks with very fine vertical specialisation by function or component between regions; and this has a major impact on technology transfer. Export activities are substituted by technology transfer flows. This trend is beneficial for local technology recipient companies, because they obtain access and absorb MNCs' technological know-how and management practices. For instance, Intel Malaysia has created a programme for local suppliers (SMART) that has five steps: select promising suppliers; provide initial training; allocate contracts according to capabilities;

raise capabilities by further technical assistance and training; and help suppliers diversify and become global suppliers (Lall 2007).

From the point of view of territories, technology transfer between advanced and developing regions is particularly important, especially if developing regions want to get fast-changing proprietary technology not available at arm's length. This relationship, referred to also as 'north–south' technology transfer, was studied by Datta and Mohtadi (2007) who described an endogenous growth model, in which technology, received through imitation and the level of human capital, determines the south's ability to move from imitation to innovation. The model shows that the south becomes a 'quasi-innovator' when it has human capital exceeding a certain threshold level. As long as the north innovates and the south imitates, the two regions diverge. The cost of imitation is negatively related to the size of the knowledge gap. Regions away from the world technology frontier have relatively smaller imitation costs; however, costs increase as the knowledge gap decreases. The north's gains are the highest when it trades with regions poor in human capital, because a human capital-poor south is more dependent on imitation and thus on the import of intermediate goods from the north. The model predicts growth convergence between the north and the south that has started to innovate.

Figure 7.1 describes a mainstream understanding of technology transfer as a process of imitation that differs from innovation. However, technology transfer is moving ever closer to innovation, as licensing of IPR demands additional elaboration before it takes the form of a new product or technology. This is true in start-ups and spin-offs and the commercial exploitation of academic R&D with the support of risk capital and incubation facilities. Patents and R&D results also need additional processing before becoming a new product or process. However, the rise of knowledge-based and innovation-led development trajectories created strong pressures to find new sources of knowledge and technology and led to a reconsideration of the roles of licensing, technology buy-out, and cooperation networks.

The turn towards academia

In the first decade of the new century, academia, research centres and university laboratories are gaining ground in technology transfer, becoming strategic partners and support organisations for technology transfer, forming pools of expertise and know-how available to entrepreneurial activities (Dosi *et al.* 1988). The reason is that technology transfer from universities can provide advanced and promising technologies, bypassing barriers that keep proprietary technologies with MNCs and other major players. In less favoured regions, in particular, with very limited or non-existent foreign investment and backward productive tissue, universities are the unique source of advanced technology and innovation.

Furthermore, the needs of enterprises for leaner production and speedy renewal of products and technologies have created a strong demand for acquiring technology from external sources, which in turn, has enhanced the role of universities and other research institutions. As discussed, the 'closed' innovation model, based on internal R&D departments developing technology in-house for the sole use of their corporate needs, is becoming more and more obsolete (Chesbrough 2003). Modern business leaders are practising 'open' innovation, in which companies import knowledge from external sources, wherever this is available, while letting their own R&D enter the market via other organisations.

The purchase of equipment and machinery as the dominant technology transfer route has become eroded in favour of inter-firm and university–industry cooperation (Table 7.1). These recent trends highlight the fact that network structures and institutional agreements take the lead, limiting the market-mediated technology transfer.

The orientation of universities towards technology transfer and exploitation marks a significant change in their structure, with them now integrating teaching, research, and the provision of services to industry. The establishment of communication channels and mechanisms for transferring technological knowledge and R&D from research institutions to companies is necessary in order to bring this knowledge into use. The physical space of the university campus also changes with the creation of cooperation spaces (science parks, incubators) and the opening of infrastructure in non-academic users.

The impact of technology transfer from universities to firms can be observed in many successful paradigms of innovation development. The emergence and development of small, high-tech firms – mostly spin-offs – from universities, in Île-de-France, Cambridge, and Milton Keynes, for example, is based on their collaboration and close links to universities located in the same area. Knowledge generation and dissemination from universities to firms through in-between close collaborations is the key factor for the development of these small innovative firms (Cooke 1996; Crang and Martin 1991). Another example is the industrial cluster that is characterised by networking and strategic alliances among geographically proximate firms, inter-related in the value chain. Clusters cannot be effective unless they form an extended network with universities, research

Table 7.1 Changing technology transfer routes (EU-15)

Technology transfer routes	May 2001 (%)	September 2002 (%)	Percentage change
Cooperation with universities and consultants	11	14	+27.27
Inter-firm cooperation	51	59	+15.69
In-house R&D	30	31	+3.33
Purchase of equipment and machinery	61	41	−32.79
Licensing of technology	12	9	−25.00
Other	4	3	−25.00

Source: Based on Flash Eurobarometer, September 2002

centres and development agencies in order to obtain a base for knowledge generation and information (Juniper 2002; Simmie and Sennett 1999).

Recent studies pointing out the public benefit and the economic impact of the technology transfer process focus on defining the different patterns by which technology dissemination from research institutions to firms takes place. From these observations, two distinct models of university technology transfer process can be identified:

- The first refers to the establishment of *formal procedures* between universities and firms through formalised research processes, such as research contracts, patents, or 'buy-sell' transactions;
- The second considers technology transfer as a *collaborative activity* occurring within networks of formal and informal relationships between universities and firms.

(Harmon *et al.* 1997)

In the former model, technology transfer is seen as a linear process resulting from a one-dimensional relationship between technology supply and demand, where suppliers and users of technology operate independently and the gap between them is bridged by a licensing contract. This process presupposes more formal technology transfer procedures in terms of contractual search or patenting the technology developed at the university and transfer of patent rights to firms. It also suggests that both business and universities' laboratories must find ways to respectively externalise and publicise their offerings. A problem arising from this model is that the majority of firms, and specifically small ones, do not have a well-developed, or even any, formal search procedure to find a technology needed (Harmon *et al.* 1997). Thus, universities should find appropriating channels to overcome the constraints in bridging technology supply to demand.

The second technology transfer model relies on interdependences, interactions and interactive learning between different bodies, arising from collaboration networks and collective learning processes (Cooke and Morgan 1998; Cooke *et al.* 1997; Edquist 1997; Freeman 1995; Lundvall 1992; Lundvall and Johnson 1994; Morgan 1997). According to this view, interactions through various forms of cooperation and networks activate learning processes, which in turn activate the acquisition of knowledge. The learning process, as a process that promotes the capacity of firms to get new knowledge, takes the form of a collective process, which occurs via various inter-firm relations and cooperation between firms and research bodies, via joint research projects, and other formal or informal links. A significant outcome of such a collaboration system is the sharing of tacit knowledge that enhances information flows and innovation. This collective sharing and transfer of knowledge among the actors involved in the process constitutes the basis of territorial systems of innovation and localised interactive learning processes (Kyrgiafini and Sefertzi 2003).

In both models of technology transfer, via formal procedures or collaboration processes, research activity is seen as a major contributor to the creation of information, knowledge and new products, which in turn needs distribution channels to arrive and be absorbed by companies. In the formal model, a major problem derives from difficulties in bridging supply and demand, because firms usually do not know what universities do, or because universities do not know what firms need. Since the transfer of technology is the movement of technology via communication channels from one organisation to another, exteriorisation of the university activities, establishment of information systems and elaboration of ways to disseminate existing R&D is of prime importance (Rogers et al. 2001). In the collaborative model, even if it seems to be a more efficient way of transferring technology, difficulties also arise regarding the establishment of collaboration networks. Obtaining an interactive learning process through close collaborations, presupposes an appropriate business culture, an institutional support framework, interface mechanisms, and trust for cooperation, knowledge sharing and collective learning. Both universities and companies fall behind in setting such frameworks.

Technology transfer as communication

Considering that technology is the knowledge, means, and ability to resolve a problem, technology transfer is the process by which this ability is transferred from a source to a user, along a communication network that connects sources and users. The characteristics of the network are determined with respect to the source, recipient, what is transmitted, context, transmission means, and facilitators. These traits influence technology flow, speed of transfer, costs, and support during the process. The network architecture differs when the transfer is from one source to one user (one-to-one), one-to-many, or many-to-one. Complexity is high, shortcomings and failures are frequent; thus the need for technology intermediary organisations, technology transfer offices and officers, brokers, patent attorneys, and other facilitators who intervene in the communication process, safeguarding agreements and knowledge transmission.

Two major functions of this communication network are the codification of information and knowledge that is transferred and its de-codification/interpretation by the user, generating the intended ability or competence (Garavelli et al. 2002). Codification is made at source, according to the cognitive system of the codifier/source. Codification may be addressed to a wide range of users, in the case of technologies embodied in capital goods, or to specific users in the case of targeted licensing. Codification enables technology to be managed via objects, such as databases, software, reports, flowcharts, demos, licensing agreements, etc. Raw data (measurements of environmental stimuli), information (giving meaning to data), and knowledge (interpretation and use of information to resolve a problem) may be codified and transferred. This ever fresh categorisation of Davenport and Prusak (1998) points out different levels of know-how that flow over technology transfer networks.

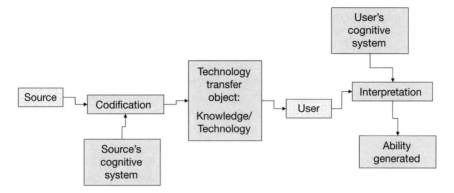

Figure 7.2 Technology transfer as communication

De-codification and interpretation of information and knowledge is made by the user and the technological ability is transferred. Interpretation is fundamental in technology transfer. It depends on the user's cognitive characteristics and the context of knowledge transmission. Only part of the abilities codified will be transferred, because of the capacity and effort of users to absorb the codified technology.

Source-user communication and exchange networks, codification and interpretation are assisted by intermediary organisations. The more technology transfer is embodied in objects the more the need for assistants in various forms (help desks, call centres, virtual assistants) helping to correctly de-codify the messages, reduce noise, and avoid misunderstanding. The same holds for culture: the greater the cultural gap between sources and users the greater the differences in the cognitive systems of interpretation and the need for physical and virtual assistants.

Blending cooperation networks and virtual spaces

To date cooperation and communication networks have marked the landscape of technology transfer. However, these networks have found a powerful ally in ICT-based networks. The rise of the information society has paved the way for new forms of technology transfer intermediaries, such as online help desks, virtual brokers, virtual technology spaces, and online assistants.

Organisations have started using communication technologies and the Internet in order to facilitate the dissemination and commercial exploitation of technologies. Two major tendencies have appeared:

- On the one hand, there is the proliferation of call centres assisting users in applying technologies embodied in commercial products. Most major MNCs cooperate with call centres and help-desks. They operate all over the world,

but Bangalore, India, has been recognised as an important location for this activity. Young, motivated employees from Karnataka are trained to help customers in various geographic locations, the US, Europe and elsewhere. They provide direct help in resolving operational problems, and answering questions related to technical and other relevant issues.
- On the other hand, numerous virtual technology transfer spaces have been created by large companies, academic and research institutions, technology brokers, and consultants. They host online databases and virtual tools and offer professional services assisting other organisations in adopting new technologies. Through such applications, companies gain access to research results from all over the world, can communicate their particular technological needs, and get insight on technologies and applications.

Both forms of ICT-aided technology transfer help bridge the geographical distance between technology providers and users, different time zones, language and other barriers. The user may contact and speak to an expert or see online a video or a demo; have a virtual assistant instead of a tutor; consult a database with FAQ and obtain the opinion of satisfied and dissatisfied users; follow a roadmap indicating stages and steps of implementation. All these facilitators are meaningful for covering the tacit knowledge aspects that exist in any technology. They substitute the need for direct person-to-person communication, and replace physical contact with digital communication and real-time interaction (Azzone and Maccarrone 1997; Cargo 2000; Juniper 2002).

A variety of virtual technology transfer spaces and web-based tools have been created to facilitate organisations and companies in finding and absorbing technology. We can classify them into three categories, according to the content and online tools that they incorporate.

The first category includes *virtual marketplaces* that offer technologies and enhance interactivity between technology and business communities. They contain the R&D intellectual property and utilise the Internet to make it widely available. Online technology marketplaces offer universities and research institutions the ability to extract tremendous value from the intellectual property they are willing to share. CORDIS Marketplace, available at www.cordis.lu/marketplace, and Yet2.com Global Technology Marketplace, available at www.yet2.com are good examples of online marketplaces for buying and selling licensable technologies, know-how, processes, and similar intellectual property.

The second category is *virtual spaces that contain online tools* based on expert knowledge. These tools can help users to solve specific problems that arise during technology transfer. An example is the *Virtual Technology Transfer Platform* available at www.newventuretools.net. It is a virtual technology transfer environment, which has been developed by a network of European technology parks, university labs, and technology transfer centres from Finland, Germany, Greece, and Portugal. Six online tools are available free of charge to help resolve typical technology transfer problems, such as technology watch, technology audit, technology assessment, networking, marketing of innovation, and financing of innovation. Virtual technology transfer assistants complement virtual teaching applications that are already successfully being utilised in universities in online training and education.

The third category is *virtual spaces that offer e-learning* in the field of technology transfer and the management of innovation. Virtual spaces enhance the interactive learning process by integrating other aspects of the learning process, beyond the delivery of information, such as demonstration, assisting companies in resolving problems, learning by reflection and assessment (Allen 1998). Virtual training is based on roadmaps and step-by-step learning. By following the roadmap's steps the user is exposed to methodologies about how to solve a problem, while support material is available (procedures, tools, companies and organisations, case studies, etc.), and additional assets (articles, presentations, sample deliverables, references, etc.). At the end, the learning exercise is concluded with exercises, and evaluation of results.

The combination of virtual technology marketplaces, online technology management tools, and e-learning applications with technology cooperation networks has given birth to a new form of technology transfer: **Augmented technology transfer**. The latter provides better functionality and communication in the sub-processes of technology transfer. Augmented technology transfer amplifies all critical aspects of technology transfer: codification, interpretation, and intermediation. On the codification side, it provides solutions to reduce the effort needed for a full description and demonstration of the technology. On the interpretation side, multimedia and step-wise assistants make understanding easier and coherent. On the intermediation side it offers direct help through online communication and assistance.

The DRC: a virtual space for the valorisation of academic research

Some of the above ideas about technology transfer, the new role of universities, virtual spaces, and networks have been tested at the Digital Research Centre (DRC) of the Aristotle University of Thessaloniki. It is an experiment targeted at the community of academic researchers wishing to cooperate with companies, transfer and implement technologies and products created in university research Labs. The Centre was set up by an academic network of ten Labs from the departments of Urban and Regional Planning, Architecture, Agriculture, Mechanical Engineering, Civil Engineering, and Informatics.

The rational behind its creation is quite straightforward. The university research community has developed a significant reserve of research results, and research-based products and services, which are not widely known, constitute remarkable applications but with limited applications in the fields of industry, services and public administration. The DRC seeks to collect this intellectual wealth, make it more widely known, and facilitate application and utilisation of it. It is a centre for transferring technologies from the university research community to a large number of external users.

184 *Building blocks of intelligent cities*

Planning of the Digital Research Centre was based on extended market research that assessed technology needs and active demand in the sectors of agriculture, manufacturing, construction, energy, consulting, insurance, transportation, information and communication technologies, banking, tourism, and health. The market research revealed the absence of an R&D department in the majority of regional companies and the absence of systematic collaboration between academic organisations and companies; the survey also helped define the areas of higher demand for technologies and technical expertise.

The DRC combines both a physical and a virtual dimension. In physical terms, it covers the university campus and the Labs located within the campus. The network of academic labs with their equipment, people, and facilities make up the core of the DRC. The virtual dimension includes a series of e-tools facilitating technology transfer and the exploitation of university expertise. In particular, the Centre consists of three layers (Figure 7.3).

The first layer is formed by an online database for storing and disseminating R&D results. The most important research outcomes, especially those that may lead to the development of new products, new production processes and new services, are listed in the database. Technology providers from universities and other research and technological institutions can submit profiles and detailed information about their research, products, and services. Technology users from both the private and public sector can access this information over the Internet. The database entries are categorised in scientific categories and market

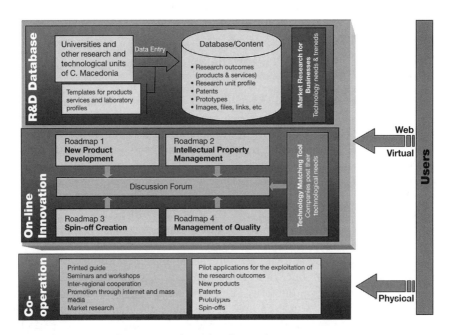

Figure 7.3 DRC architecture – physical and virtual components

applications so that companies and public organisations can easily find what they are searching for.

Data collection is based on three registration templates that each research organisation has to complete. The first template describes the organisation; the second the research-based product; and the third the research-based service. Researchers, labs, and technology providers can fill out these templates online or by e-mail. After validation and verification, these are stored in the database. The type of information describing research products and services was determined after in-depth discussion within the research community. The purpose is to provide adequate information while not overloading the registration procedure. Market research and companies' views have also been taken into account. Technologies are categorised using two different classifications:

- According to scientific categories and academic disciplines to which the technology belongs. The selected scientific categories and subcategories satisfy the needs of academic organisations primarily, and provide information about the fields of science and technology related to the products/ services in question.
- According to market applications and industrial sectors in which the R&D may have a potential use. Available R&D results, technologies, and products are classified with respect to their relevance to different industry sectors, including manufacturing sectors, energy, construction, environmental, information technologies, automation, quality assurance, business services, and other NACE categories. Each of the above market application areas has several subcategories.

The second layer is formed by an online technology transfer platform. It is based on online roadmaps that clarify aspects of R&D exploitation and use. The roadmaps are complete methodological guides that help users (laboratories, companies, technology brokers, and intermediary organisations) accomplish tasks related to new product development, spin-off creation, intellectual property management, and management of quality. Furthermore, this layer includes two online communication tools between academia and businesses, a technology-matching tool and a discussion forum. Both create a digital space where entrepreneurs, SMEs, and public organisations can post their technology needs that are automatically communicated to the closest lab or technology provider in order to open a dialogue and find a solution.

Four virtual assistants/roadmaps have been created:

- The new product development roadmap is a learning platform and collection of tools that guides the user through the five stages of new product development: ideas generation, assessment and screening of ideas, product design, prototyping, and commercialisation. At each stage, assessment points allow the process to be continued or killed.

- Spin-off creation guides the user through the four stages of a spin-off company's creation process: identification of a potential new product, analysis of the business opportunity, investigation of issues regarding intangible assets, and preparation of the business plan. Assessment procedures link the different stages.
- Intellectual property management is a roadmap for the creation, management and commercial exploitation of intellectual property. It covers various topics related to intellectual property such as licensing, patents, copyright, disclosures, trademarks, registered designs, design rights, etc., while also providing the relevant legislation and case-law.

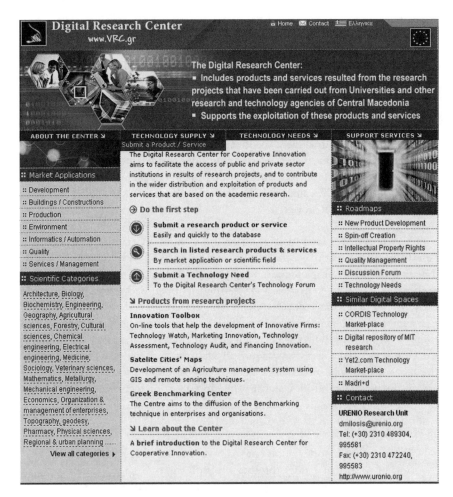

Figure 7.4 DRC layer 1 – a database of technologies and expertise
Source: www.vrc.gr/index_en.html

- Management of quality helps to introduce and operate quality control systems. It informs users about the international standards and presents quality management and certification procedures. Here the user may also find accreditation of testing and calibration laboratories. The roadmap is particularly important for laboratories working in the field of testing and destructive evaluation and for companies searching for such services.

The roadmaps are divided into thematic steps. Each step deals with a specific problem and provides methodologies and tools for solving this problem. At the disposal of the user is support material, tools, demos, directories of companies and organisations to consult, case studies, articles, reports, references, etc. At the end of each step the user should complete a deliverable and evaluate it against a given self-assessment template (Figure 7.5). Where appropriate, the user can use other online tools available over the Internet.

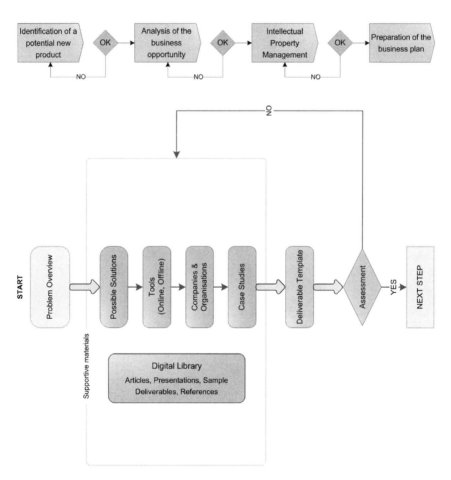

Figure 7.5 DRC layer 2 – structure of spin-off creation roadmap

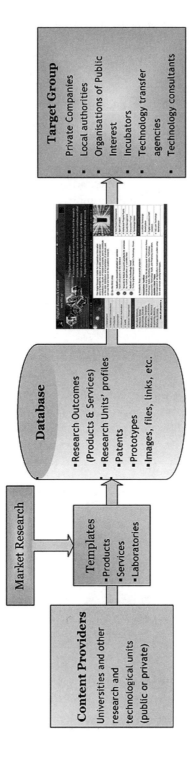

Figure 7.6 DRC layer 3 – technology dissemination

The third layer covers technology dissemination activities and small pilot projects testing technology transfer, mainly the co-financing of efforts where companies and university laboratories cooperate in the development of commercial products or services based on R&D results. A series of technologies were supported financially to make their commercial exploitation easier, including: evaluation of the use of high resolution satellite images in identification and mapping of cotton fields; development of software for calculating long-term levels of atmospheric pollution in urban areas; development of software to simulate atmospheric flux and pollution dispersion at local level; digital information systems and databases for metropolitan areas; development of software to assess seismic movement in the design of public constructions and buildings; management of agro-environments with GIS applications; digital platform for monitoring innovation and regional development; energy saving in agriculture with the use of wind turbines; modelling and prototyping glass greenhouses. Most of these activities take place within the physical space of the Campus where the labs of the DRC are located. Person-to-person communication, seminars, and technology days are the main tools of dissemination.

The operation of the Digital Research Centre depends strongly on the mobilisation and active participation of researchers and labs. A network of research, production, and technology transfer agencies has been formed in order to utilise the services of the DRC: procurement of knowledge and information about 'who is doing what' and 'who needs what' in various fields of research and technology. Overall the DRC seeks to promote the exploitation of technologies and research conducted in university laboratories, recording the technological needs of the regional companies, and cross-linking technology supply and demand.

The impact of the DRC on the regional economy is expected in three domains. First, by improving the access of firms to scientific research. Research in the region is mainly conducted by university labs and, although it covers a wide range of scientific areas, it is somehow far from the needs of regional firms. Existing links and cooperation between research and production are quite limited, mainly due to the isolation of both parts and the lack of interaction mechanisms. The DRC is being developed so as to bridge this gap, enabling the commercialisation of work done in laboratories. The centre provides the mechanism for bringing the technology supply side (university laboratories, research and technology transfer institutions) closer to the demand side (companies). Second, by improving the capabilities of technology transfer organisations, systematising technology offer and demand, having online assistants, and better understanding the needs of their customer base. Third, by fostering an innovative culture in the region. The easy accessibility of the DRC, the large number of users of its services – companies and research laboratories – allow a wide dissemination of technologies, and make technology transfer more familiar and accessible to all.

The first two years of operation of the DRC revealed a series of difficulties to be overcome and shortcomings in its functioning, both from the perspective of the technology providers who list services and technologies in the virtual space, and from the perspective of users who visit the virtual space and use its tools and online applications.

From the technology provider's point of view the effectiveness of an online R&D dissemination platform is directly dependent on its diversity and richness in terms of R&D content, technologies, and products. The virtual technology transfer space of the DRC has to be developed in close cooperation with R&D labs and technology providers. It is not possible to construct such a database without the direct involvement of the R&D organisations, research teams, and technology developers that took part in R&D projects and produced the technologies. It is not a question of intellectual property rights, though this dimension also exists. Only R&D and technology organisations have the knowledge to precisely describe research outcomes and provide information about the technology in question, its eventual use, sectors of application, and conditions for exploitation.

In order to facilitate cooperation with R&D organisations and technology developers, we created different templates to codify the collection of information. As mentioned, there are three templates available, one for labs, one for products, and one for services. The provider just has to download the template from the DRC website, fill it out, and send it back to the DRC. The simplicity of the procedure aside, we found considerable difficulties in getting R&D and technology providers involved. The problem behind this reluctance is lack of motivation and low expectations for technology exploitation. R&D organisations do not seem sufficiently convinced about the value of online technology dissemination, negotiation, and transfer; in other words convinced that there will be a return from registering R&D results in a public database.

Interviews on the causes of this reluctance and the motivation of technology providers for listing R&D information revealed three main obstacles. The first is organisational; you have to always go down to the research teams in order to get accurate information. Even in large organisations, this information is rarely stored in advance or transferred to administrative personnel. The second relates to the volatility of research teams; it is very often the case that the research team is dissolved at the end of the research project. This diminishes interest in the post-project phases and eventual exploitation of R&D. Chaotic intellectual property right settings also decrease the involvement of the initial R&D team in post-project exploitation. The third obstacle concerns the maturity of the online technology marketplace itself; to date very limited online technology transfer has been taking place, and few users are looking systematically for technology via this route.

The meaning of these obstacles is that virtual spaces may easily lose their collaborative dimension. The permanent link to research teams and organisations is crucial. The virtual space is less a repository of past knowledge than a tool enabling one to follow the entire lifecycle of research, from conception, creation, and application of knowledge. A pure virtual function impoverishes the full potential of online innovation and technology dissemination.

From the user's point of view, the real added value of online innovation and technology transfer platforms and tools is tied into its openness. Interviewing users about technology dissemination through the online platforms of the DRC, we observed a very positive appraisal of the opening of university R&D up to public eyes. This was emphatically stated by small innovative companies working with internal product development teams. For them information about R&D from university labs is an additional source of inspiration and product innovation. The non-profit character of universities and the tradition against information disclosure helps online applications to be viewed as sources for new ideas, products and technologies from which they could benefit.

Investigation about the use of the virtual space in technology licensing revealed two main domains of interest. The majority of users visit the virtual space to learn about third party R&D and find technologies that match their specific technology needs; a smaller percentage looks at these spaces to better understand technology licensing and find models and best practice on formal technology agreements and licensing contracts. However, users find regional R&D information systems more appropriate than global ones (such as Cordis or Yet2.com), which is explained by the fact that the former offer information in their language. Another advantage of regional technology databases is when a technology is found, it is easier to get in touch with a provider located nearby and take additional information or receive a demonstration. Cultural (linguistic) proximity and geographic proximity make communication and technology cooperation much easier.

A critique frequently addressed concerns the summary presentation of technologies or R&D results, which makes understanding them rather difficult. This clearly refers to the question of interpretation of technologies embodied in objects. No doubt, multimedia presentations, drawings, pictures, and demos may improve the description of content and understanding; but these forms of communication need more time spent on preparing and filling out the templates, which is not always feasible. This critique neglects the fact that the virtual space is just the first step, for finding a reference point for co-operation; next steps rely on direct communication between technology providers and users.

In the field of technology learning, online roadmaps were very much appreciated. Demonstration of the new product development roadmap in higher education engineering departments revealed an interest in using it in the classroom for computer-assisted product development. In all cases, knowledge disseminated through these applications is more formal than tacit knowledge. Procedures and expected results are clearly defined in advance, though the way these tools are used is open to imagination and creativity.

It also became clear that taking advantage of online R&D information and technology management tools is only possible when companies have some internal R&D or product development capability. At least a small in-house product development team is necessary to adapt external research and technology to the company's needs for products and technologies. Organisations lacking in internal R&D and in-house technology development capability cannot take advantage of online applications. Adoption of new technology and innovation is possible by combining internal and external capabilities. A fully distributed innovation model is not operational.

Understanding the added value of online technology dissemination and problem-solving suites allows their effectiveness to be optimised. Most important in our view is the articulation between institutional networks and virtual technology transfer spaces: the use of online platforms and tools to assist the activities of technology intermediaries, and innovation teams. The intelligence of digital innovation spaces may be gradually improved from this link to experts and R&D labs and teams. Having permanent feedback from researchers, technology experts, and innovation professionals enables the internal procedures and knowledge generation functions of digital tools to be trimmed.

The information technologies used are rather conventional; most design and development creativity goes to the setting of logical circuits and knowledge routes, allowing a complex problem to be broken down into its simplest constituent components. The conceptual framework on which online tools rely, combines the codification of expert knowledge, the follow-up of routines, and increased information storage and retrieval capacities.

Currently two universities in Greece are using this platform to better exploit their R&D intellectual property. The Research Committee of the Aristotle University of Thessaloniki operates the Digital Research Centre (www.vrc.gr), and the Liaison Office of the Democritus University of Thrace operates the Innovation Management Digital Space (http://innovation.duth.gr).

Augmented technology transfer

Technology transfer is the second pillar of intelligent communities and cities. A developed technology transfer system enables cities and regions to find and assimilate state-of-the-art technologies wherever they are available in the world. However, this ability does not automatically emerge. It needs to be organised and cultivated.

The technology transfer tissue of an area is comprised of numerous organisations and institution: technology transfer centres, incubators, innovation centres, university liaison offices, consulting companies, innovation experts, technology networks and clubs, and larger facilities like science and technology parks. Virtual technology transfer spaces and tools augment this tangible tissue and improve its operation. Combining physical and virtual technology transfer spaces, cities and regions obtain a more effective technology transfer system. The functionalities that the virtual space adds to technology transfer centres and networks differ from case to case. However, a common trend of augmented technology transfer is the advancement of a series of capabilities related to technology intermediaries, such as communication, networking, technology offering, involvement of users in problem solving, and service delivery.

Deepening of technology transfer is the main characteristic of augmented technology transfer. Improved capabilities and virtual spaces are necessary to sustain the current orientation of technology transfer from horizontal to vertical technologies.

The technologies transferred and absorbed in an area can be divided into horizontal and vertical ones. Horizontal technologies are those that apply to all sectors in a region. They are defined by their technological content rather than by their scope. Characteristic examples are energy saving technologies, production organisation and optimisation technologies, automation, quality management, clean environmental technologies. On the contrary, vertical technologies are those that are useful in specific industry sectors, such as food processing technologies, chemical technologies, steel and metallurgy, plastics and rubber processing technologies. This distinction is normal. For example, it is found in the EU technology platforms, which have developed cooperation networks and set targets either per technology without reference to the production sectors involved (plants for the future, photovoltaics, sustainable chemistry, mobile and wireless communications, innovative medicines) or by technology suitable for specific production sectors (steel, textiles, construction).

In September 2006 we conducted a survey in technology transfer centres across Europe, as part of the design and development of a new technology transfer centre (URENIO 2007b). We examined the services offered by a large number of centres and we came across a very interesting discovery: that most centres focus on horizontal rather than vertical technologies. The reason is quite understandable. In the case of horizontal technologies, the advantage is that the same technology transfer agency can cover all industries, ensuring major economies of scale. On the contrary, in vertical technologies different technology units/teams are necessary because the services provided differ significantly. The advantage of scale is lost. Due to the fragmentation of demand, most technology transfer centres place emphasis on horizontal technologies or a combination of horizontal and a few vertical ones. For each horizontal technology, a joint technology transfer strategy can cover all processing sectors. On the contrary, in vertical technologies, specific strategies and skills are needed in each industrial sector.

Augmented technology transfer and online marketplaces are changing this landscape in a substantial manner. They enable different technology supply architectures to emerge, based on networks, cooperation, and specialisation. As a result, newer technology transfer centres give higher priority to vertical rather than horizontal technologies. In fact, the virtual space offers two additional functionalities sustaining this trend:

- better networking with technology suppliers, multinationals, universities, and research institutes from every region of the world, which is translated in a wider portfolio of technology offers; and
- better ability to interpret technology offers, which means better ability to understand, describe, demonstrate, and evaluate the technologies offered.

Better networking marks the quantitative aspect of augmented technology transfer, as it widens the opportunities for acquiring and transferring technologies. Applications such as the DRC contribute to expanding the scope of potential technologies that can be adopted and used. Better interpretation marks the qualitative dimension of technology transfer, the capacity to describe better a technology, to foresee alternative opportunities for technology application, based on virtual assistants, better demonstration, and virtual technology assessment tools. Both functionalities are important for organisations offering technologies and brokers who promise to offer the most effective technology from every corner of the world. They make the world more open to cooperation.

8 Innovation through collaboration

Managing networks that cross boundaries

Innovation through collaboration

Over the last 15 years we have witnessed a profound shift in the theory of innovation. Research from both sides of the Atlantic has shown that the processes of innovation have changed radically. Novel ideas for technologies and products, which traditionally emerged from R&D labs internal to the company, now spring from networks of collaboration between companies, universities, suppliers, and customers. Chesbrough (2003) described this change as an evolution from a 'closed' to an 'open' innovation model, imposed by new knowledge landscapes in which external knowledge sources become more important than internal ones.

Numerous studies have confirmed the trend towards external innovation partnerships and outsourcing (Arora *et al.* 2002; Quinn 2000; R&D Magazine 2002). The same trends have been observed in small and medium enterprises (Macpherson 1997) as well as in large multinational firms (Love and Roper 2004). In the UK, for instance, extra-mural R&D doubled in the period 1985–1995 and its share in total business R&D expenditure grew from 5.5 per cent to 10 per cent (Howells 1999). The MERI-CATI database, which primarily covers large firms, reveals a major increase in the number of inter-firm innovation partnerships and in collaboration on an international scale (Caloghirou *et al.* 2004; Hagedoorn *et al.* 2000). The study by Goldense and Schwartz (2002) on large companies in Europe, the US and Japan revealed that 90 per cent outsourced some part of their R&D activity and around 95 per cent of this outsourcing focused on the development of new products. Roberts (cited in Kimzey and Kurokawa 2002) concludes that the most important change by companies around the world is the 'relentless intensification' of dependence on external resources for technology. Although he reports significant variations among firms in the US, Europe and Japan, there is a clearly established common positive trend. Increasing demand has also been accompanied by the development of a new market in product development services by specialised, private companies and public research and technology organisations (Chiesa *et al.* 2004).

These innovation and new product development networks are extremely complex. They differ from technology transfer and absorbing state-of-the-art technology in terms of focus (having a specific product to make), risk (there is no guarantee that the product will meet technical specifications and market demand), and novelty (having a new product to make). To deal with these challenges innovative companies are developing networks with three different loci, intensifying cooperation with academia, other companies, and customers:

- *Partnerships with academia* have become a strategic component of collaboration. Companies tap into knowledge found in universities and in many cases localisation decisions are driven by the prospect of such collaboration. Partnerships with academia are a core objective in many national R&D policies, and US and EU science and technology policies as well.
- *Inter-firm strategic technology collaborations* are more and more frequently used as a source of knowledge. Joint R&D is used to complement internal resources in the innovation process, enhancing innovation input and output. The intensity of in-house R&D also stimulates the probability and the number of joint R&D activities with other firms and institutions (Archibugi and Coco 2004; Becker and Dietz 2004).
- *Customers* are asked to express their needs and suggest products and services. Complaints and ideas are routed from front-line customer contacts, people in sales and marketing departments, to R&D and product design teams. New product ideas are no longer developed in an isolated research laboratory, and customer input is important in shaping product development processes.

Such innovation networks, in the form of vertical supply chains or horizontal collaborative links, are spreading out on a global scale. International R&D alliances have become more and more important since the mid 1980s, and large companies are involved in multiple R&D and technology alliances simultaneously (Lichtenthaler and Lichtenthaler 2004). Firms do not seem to constrain themselves in cooperation with one single partner; in most cases they develop selective partnerships at different stages of the process (Blomqvist *et al.* 2004). Product and process innovations incorporate knowledge and components from around the world and the innovation process transcends local clusters and national boundaries. Santos, Doz, and Williamson (2004) point out that the globalisation of innovation processes is taking place in three consecutive steps: (1) prospecting (finding relevant pockets of knowledge from around the world); (2) assessing (deciding on the optimal 'footprint' for a particular innovation); and (3) mobilising (using cost-effective mechanisms to move distant knowledge without degrading it).

The logic of collaborative innovation networks

Collaborative innovation networks may be described from different points of view, with respect to the objectives behind networking, the actors that take part in it, the connectivity among members, and the degree of autonomy/ dependence among participants.

Hagedoorn *et al.* (2000) use the term 'network' to refer exclusively to close-to-market relations, in contrast to strategic alliances that have longer-term objectives. Freeman (1991), on the contrary, uses the term as an all-encompassing term covering the whole range of possible external relations among multiple players/partners, involved in outsourcing, strategic cooperation, and other alliances. The second wider meaning better reflects actual innovation networking.

Objectives

Networks exert influence on their actors and shape their activity. Networks represent both a structure of relationships among partners as well as a process of building relationships (Tidd 1995). Networks denote the presence of multiple partners – rather than bilateral partnerships – connected among each other. Issues of position in the network – central or peripheral – and power are considered critical (Tidd *et al.* 2001). As a system of relationships, the network is more than the aggregation of its constitutive parts.

There is a series of reasons why innovation networks form. Pittaway *et al.* (2004) analyse in detail recent literature on innovation networks and enumerate six reasons for setting up innovation networks: (1) risk sharing; (2) obtaining access to new markets and technologies; (3) speeding products to market; (4) pooling complementary skills; (5) safeguarding property rights when complete or contingent contracts are not possible; and (6) acting as a key vehicle for obtaining access to external knowledge. Furthermore they state a series of reasons for setting vertical new product development networks, among others: to have a significant impact on cost, quality, technology, speed and responsiveness of buying companies; to help producers identify improvements that are necessary for them to remain competitive; to enable firms to bring to bear wider expertise during the development process; and to give easier access to supplier knowledge and expertise in the longer-term.

Actors

In December 2005, IBM hosted a conference on collaborative innovation aimed at corporate chief technology officers. The speakers were analysts, academics, and industry guests from leading companies, who argued that companies are rethinking their approach to research and development in order to meet the pressures of bringing new products to market in today's global market place. The message of the conference was that companies collaborate to innovate. Traditionally companies relied on the internal 'do it yourself' model. Success came from command and control, managing innovation, assets and capabilities; direction came from within. In this model, IP means 'Intellectual Protection'. Now more and more people are embracing a new collaborative model. They have started developing symbiotic relationships, creating win-win partnerships, and utilising multi-company networks. It is an evolutionary process starting with buying the 'best from anywhere', moving to 'nothing invented here' and ultimately, to sharing assets and expanding. Within this model, IP means 'Intellectual Partnering'.

Radjou (2004) described this model of collaborative innovation as a network composed of four types of organisations/actors: (1) inventors; (2) transformers; (3) financiers; and (4) brokers. Inventors are creative agents who conduct basic research and design new products and services resulting in the patenting of inventions. Transformers are multifunction production and marketing services that convert inputs from inventors into new products and sell them to their internal or external customers. Financiers fund inventors and transformers and seek to own intellectual property rights for inventions. Brokers, finally, are market makers who find and connect service providers with the network, buying or selling services, and enriching the capabilities of the network. These four specialist roles provide the services that make up innovation. Rather than specialise, firms have to juggle and mix these roles using their internal and external resources.

Such innovation networks accelerate the invention-to-innovation cycle and increase the probability of success. Their added value is that they bring strong external knowledge and skills from users, customers, other companies, R&D institutes, etc., scattered on regional and national scale into the innovation process. However, they also revitalise business practices, changing the 'not invented

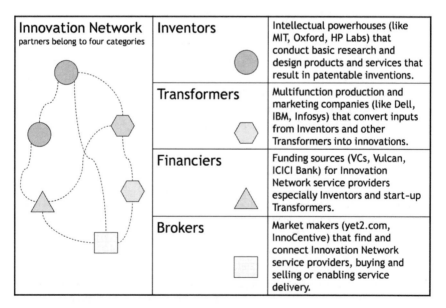

Figure 8.1 Innovation network actors and roles
Source: Adapted from Radjou (2004)

here' ethos to 'get best available expertise from anywhere'; passive customers to active innovators; economies of scale to economies of scope; IP protection to IP sharing.

Competences

Related to actors are competences that are linked. Innovation networks may connect competences from various fields: R&D (research alliances, collaborative research, strategic partnerships), technology transfer (licensing agreements, patent exploitation agreements), and new product development (cooperative product development, virtual product innovation, product development subcontracting, etc.). In each case different skills and competences are combined: R&D skills, product conception skills, prototype production, marketing and distribution skills, etc. Thus, many alternative combinations of innovation competences are possible, involving practices from R&D, technology brokering, production, marketing, and other fields. Equally important to the actors involved are the competences and skills that are gathered within the network.

Partner autonomy/dependence

This characteristic denotes the degree of dependence and hierarchy between network members. Narula and Hagedoorn (1999) classify technology networks into eight categories, which correspond to different scopes of cooperation and degrees of organisational integration, linking completely interdependent organisations to completely independent organisations (Table 8.1).

Autonomy/dependence reflect different levels of trust between the network members. Forms of dependency assure long-term cooperation, which is a precondition for building trust. Strategic alliances, equity participation and staff exchanges are management strategies to generate forms of dependence among independent organisations.

Connectivity

Innovation networks differ with respect to the way actors are connected (linear, consequential, parallel, hierarchical, mixed method, etc.). Two fundamentally different forms of connectivity appear in vertical and horizontal networks.

Vertical networks concern supplier–producer relationships, while horizontal networks are about the common use of resources. In vertical innovation networks suppliers are more involved in the later stages of product design where product manufacturability and assembly are determined, in product prototyping and testing, and in the production/product launch phase. Information exchanged with suppliers can be either at a technical level – concerning the design and technical specifications of a component or the total new product or operational – concerning availability of components, inventory and components price (Bobrowski 2000). Supplier literature (Culley 1999) can be used as a source indicating what

Table 8.1 Type of networks and degree of member integration

Mode of cooperation	Organisational interdependence
1 Wholly owned subsidiary	Completely interdependent organisation
2 Equity-based agreements (Research organisations or joint ventures)	
3 Lesser equity-based agreements (Minority holdings or cross holdings)	
4 Joint R&D agreements (Joint research pact or joint development agreement)	
5 Customer-supplier relations (R&D contracts or co-production contract or co-makership contract)	Increasing interdependence Increasing internalisation
6 Bilateral technology flows (Cross-licensing, technology sharing or mutual second sourcing)	
7 Unilateral technology flows (Second-sourcing agreements or licensing)	
8 Arms-length agreements	External transaction: completely independent organisations

Source: Narula and Hagedoorn 1999

is available off-the-shelf, checking the properties of some form, initiating information searches or sparking ideas – thus covering all phases of the NPD process. The Japanese 'keiretsu' system of partnership with the development of long-term partnerships with a few preferred suppliers has been considered by many as the model for successful supplier involvement (Bonaccorsi and Lipparini 1994).

Horizontal forms of partnerships and star alliances are primarily developed at the initial stages of pre-competitive basic and applied R&D where issues of uncompetitive market behaviour do not arise, but possibly also in areas where the firms are not afraid of spillovers and loss of core/critical knowledge (Miotti and Sachwald 2003). These interactions can be used to develop standards and common technology platforms by combining the complementary resources of firms. Firms may find areas where their cooperation can be mutually beneficial for developing new products in some areas while maintaining competitor status in others. Besides these active collaboration forms, firms can also acquire useful input in the NPD process from their competitors through fairs and trade shows, reverse engineering, competitive intelligence and other tools/methods used to analyse competitor products and develop ideas and concepts for own new products (Tether 2002).

Selection of partners: trust vs. competence

Many of these features affect the selection of partners in an innovation network. However, given the diversity involved, it is not feasible to lay down a uniform system of criteria for selecting network nodes: for selecting the partners and competences which are interconnected, their connectivity, their degree of integration and roles of each partner. It is reasonable to expect a different rationale for selecting partners depending on the objectives underlying network establishment. For example, a network to develop a radically new product will attribute greater significance to the technological skill of the partners and to protecting the intellectual property represented by the technology than to the related cost of outsourcing the manufacture of individual product components. On the contrary, in cases of limited innovation or a short copy lead time, the selection of partners offering competitive costs is a priority so as to ensure competitive capabilities under conditions where intellectual property rights are less protected.

> The type of partner firms engaged in networking appears to be related to the type of innovation occurring (Freel 2003; Kash and Rycroft 2000 and 2002). For example, incremental innovators rely more frequently on their customers as innovation partners (Biemans 1991), whereas firms that have products new to a market are more likely to collaborate with suppliers and consultants (Baiman *et al.* 2002; Ragatz *et al.* 1997). More advanced innovators, and the development of more radical innovations, demand more interaction with universities (Hausler *et al.* 1994; Liyanage 1995).
>
> (Pittaway *et al.* 2004, p. 150)

Nonetheless, there are factors systematically taken into account in selecting the nodes in an innovation network. Millson and Raj (1996) suggest that innovation managers consider four main attribute categories that influence decisions on network formation:

- the relative size of partners in terms of physical factors such as capital, employees, assets, location;
- competencies and technological factors, product lines, process capabilities, patents, R&D expertise;
- marketing factors that include the partners' existing distribution networks, customer knowledge and image and the added value in the project; and
- existing alliances of partners with other competitors, customers and suppliers, in order to diminish the danger of knowledge spillovers to other firms and loss of competitive advantage, and the possibility of the partner becoming a direct competitor in the development of the specific product.

They also suggest that the level of scrutiny and analysis and the importance of the different criteria are dependent on the type of activity that each partner is expected to perform. In the case of standardised outsourcing of well-defined contracts with low added value, physical factors and capacity to deliver the requested task according to specification will be more important. In more critical activities related to new product development and in longer-term partnerships, complementary knowledge and skills that can be transferred to the firm and marketing factors can play a much greater role.

Of all the factors influencing network formation and partner selection, we insist that two are of determinative importance in allowing long-term, productive collaboration to emerge among participants: trust and complementary competences.

Trust is the glue that binds the network together. It replaces hierarchical and dependence-based relationships that require top down collaboration. In trust-based relationships, networking relationships emerge bottom up as long-term, gradually emerging branches and connections. Networking institutions (incubators, associations, clusters) replace the lack of trust and allow a third party organisation to guarantee reliability (Cooke and Wills 1999; Keeble *et al.* 1998). The lack of trust is also complemented by technology transfer institutions and contract-based agreements. However, that is not always feasible or productive.

Complementary competences are the most important driving force in networking for innovation. Even with high degrees of trust between partners, a network will not form if complementary roles and competences are absent. A key requirement for generating innovation is the need to manage – and capacity for managing – unrelated areas of knowledge. Knowledge is dynamic and evolving. It is incorporated into products and services. It is not limited to a process of transmission but is generated, reproduced and used by all organisations involved in NPD. Regardless of which level of analysis is chosen, from lab research to modern technology clusters, one can see that innovation, diversity and complementarity of knowledge, skills and specialisations go hand in hand.

The two principles 'trust between network members' and 'competence of members' hold pole position. The solution to this antithesis takes an institutional form. Where there is no pre-existing trust this is ensured via institutional agreements. In practical terms, this means that network members are selected based on the role complementarity criterion whereas trust between them is achieved by complying with rules, institutional agreements, and regulatory mechanisms.

Spatiality of collaboration: contact vs. cost

An innovation network is a gathering of resources but at the same time an institutional relationship: an agreement between partners with different skills and roles to cooperate. Cooperation rests both on informal trust-based relationships and on formal agreements contained in contracts and partnership agreements. Moreover, networks should be beneficial for all, despite the divergent expectations and benefits for each partner.

Suppose that a typical innovation network has the form illustrated in Figure 8.2, in other words it consists of research, new product development, financing organisations, producers, suppliers, merchants and customers. Cooperation between

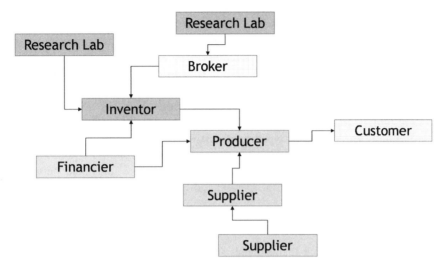

Figure 8.2 Nodes within an innovation network

the partners seeks to ensure product and manufacturing innovation, in parallel with commercial success and a match to consumer needs.

The geographical dimension of such a network is not predetermined. In the widest sense, such a network can be global, uniting partners from the remote corners of the planet. At the other end of the scale, the network can be local and all partners are located at the same place. A more likely scenario is that the network will combine local research, planning and production activities with global trade, supply and consumer aspects.

However, what would happen if the network space began to change: either by shrinking its geographical dimension or by extending it? Would all the partners congregate in the same location or scatter to the four corners of the globe? What consequences would this geographical change, spread or agglomeration have on cooperative relationships? In other words what would happen if an innovation network is transformed into a physical local cluster or a virtual global cluster?

One might reasonably expect that a change in the geographical spread of network nodes would affect the extent of communication between partners, and the cost of running network nodes. The more limited the geographical area for collaboration is, the more the width of communication and running costs for network nodes increases. The reverse also applies.

Theories of innovation are particularly enlightening on this point. Due to the tacit knowledge that feeds innovation, the dynamic of innovation collapses as the network moves beyond a certain point of spatial distribution of nodes, which negates the flow of tacit knowledge. The network ceases to function as an innovation network and falls into an outsourcing networking.

The two principles 'contact and communication' and 'network mode running costs' also hold pole position. Maintaining direct contact and extensive communication leads to all partners in the network congregating in the same location, normally entailing high operating costs, because outsourcing is limited. Vice-versa, the more network nodes are decentralised to peripheral regions, the more running costs are reduced. That is up to a point though, when decentralisation does away with the cooperation network.

Solutions lie in the combination of geographical dispersion with intensifying communication. Virtual networks may partly solve this problem, contributing to an increase in the width of communication. Through more frequent, more direct communication, even if digital, there is a higher degree of contact between partners established in low-cost operating zones.

Digital spatiality and networking

Developments in digital technologies and online cooperation platforms have affected collaborative innovation and partially resolved the above described pole positions: trust vs. competence, and contact vs. cost. Two main areas of the impact of ICTs concern their contribution to increasing the competence of partners, and in facilitating interaction, exchange and communication among partners during the innovation process, thus reducing the negative effects of spatial dispersion and distance.

Howells (1999) argues that developments in instrumentation and management of R&D activity (robotics, advanced IT systems) and the use of advanced computer aided modelling and design (CAD) applications (such as DfM, DfA, 3D-prototypes, Virtual Reality and Virtual Prototyping) favour the decision to outsource in most NPD stages. These advances led to the standardisation of more complicated processes in NPD, such as the clinical testing of drugs or building of virtual chemicals through software applications, and codification of R&D activities. As a result, there has been a reduction of tacit non-easily-communicated knowledge and a decrease of the transaction costs related with the externalising of activities. Firms can now outsource greater parts of the NPD process and much more effectively protect/enforce their intellectual property rights in relation to their partners.

Simultaneously, the development of advanced telecommunications systems and ICT applications (Electronic Data Interchange, e-mail and computer coordinated faxes, databases, etc.) ease communication among supply/service firms and manufacturers (Crow 2005) and reduce coordination – and transaction – costs. Ozer (2003) has analysed the potential role of the Internet in supporting the NPD process across its various stages. At basic levels of utilisation, the Internet can improve the availability and accessibility of information to possible partners, facilitate speedy transfers and reduce the costs of interaction. It can also be a good source of new product ideas. At more advanced levels, web-based product design and development environments can be used to integrate different functions and stages of the NPD process and real-time interaction among the different partners. One such example is the development of Virtual Users' Communities for greater participation of customers in the NPD process (Von Hippel 2002). Community members are provided with web-based tools to participate in the idea generation, as well as in product design and testing, stages of the process. Kessler (2003) describes a wide range of business practices/case-studies of Internet use as a medium for improving the speed, efficiency and quality of the R&D process. R&D networking tools are referred to as well suited for monitoring external knowledge sources and collecting information and ideas and for participating in geographically-dispersed cooperative arrangements.

While there are still important barriers in the widespread application of ICT-based cooperative innovation networks and physical proximity is still very critical in the partnership process, the development of online cooperation tools can seriously cut coordination and management costs for partnerships and facilitate interactions across regional and national borders (Sethi *et al.* 2003).

Managing collaborative innovation networks: three experiments

Below we will examine three different strategies for setting up collaborative innovation networks at global, regional and sectoral level. We will outline the forms of collaboration and examine how network formation relationships change depending on geographical scale, technologies, and collaboration agreements. However, these three examples have one important thing in common: they are networks that combine human skills, institutional collaboration agreements and virtual collaboration and knowledge management spaces. To different degrees and in differing ways they create forms of intelligent spaces.

The purpose of this analysis is to understand how complementarity emerges between members of the innovation network (nodes) and how the change in the network's geographical space from global, to regional, and inter-regional alters the skills contained within the network, trust-based relationships and the strategy of developing innovation.

When it comes to the dynamic of spatial agglomeration or decentralisation, the question is which nodes in the innovation network will remain in high cost zones and which will move to peripheral regions?

InnoCentive: innovation based on global networks

A brilliant solution for outsourcing scientific and technological research was created in 2001 by Darren J. Carroll, founder of InnoCentive. Carroll pioneered the application of web tools and open innovation principles to resolve R&D problems in the pharmaceutical industry, and in chemical and biological research in particular. Before founding InnoCentive, he was an attorney at Eli Lilly & Co. with legal experience in licensing and outsourcing transactions, mergers, acquisitions, and the commercialisation of technology (The Wall Street Transcript 2006).

The InnoCentive model is simple and clear. It is organised as a virtual innovation network facilitating collaboration between companies, contract research organisations, university labs, and freelance scientists. InnoCentive uses the Internet to connect research-driven companies to top scientists worldwide. *Seeker* companies anonymously announce scientific problems (as Challenges) and the award that they intend to pay in the case of resolution. *Solver* scientists are registered scientists and Labs that attempt to provide a solution to posted problems. *InnoCentive* is the broker that facilitates the collaboration process, defines the intellectual property rights relating to collaboration, and guarantees the award payment.

This model is aimed at large companies or smaller ones that do not have internal capacity or expertise to solve particular problems. Creating a global network of distributed research, InnoCentive complements internal R&D with research capabilities scattered all over the world. Scientists are informed about problems that major companies are facing. Companies may tap into skills and expertise of a retired head of R&D department, a professor in a remote university, a researcher in a public R&D lab, and others. The Seeker gets a valuable solution and the Solver scientist gets the opportunity to work on some interesting problem and gain a financial reward out of it.

A virtual space facilitates cooperation (www.innocentive.com). The space supports binary research collaboration relationships: one Seeker to one Solver or one Seeker to many Solvers. Additionally, it contains all useful information to start working on a posted problem. The working language is English to ensure that everyone works on the same problem definition and description. This space offers a virtual laboratory in which skills and expertise from the US, Europe, and Asia converge.

Seeker companies post 'Challenges' to the InnoCentive virtual space. Companies sign a contract with InnoCentive to become Seekers and post Challenges to the virtual space. Each Challenge (i.e. measurement of pyrophosphate, breast cancer risk assessment, DNA separation, etc.) contains a detailed description and specifications of the solution, a deadline, and the reward to be paid for the best solution.

There are two types of challenge: theoretical papers and laboratory challenges whose results need to be verified in a laboratory. Theoretical problems are mainly addressed to university staff, faculty, post-docs and students. Applied problems concern laboratories with the ability to test and measure a solution.

InnoCentive started with only one client company, Eli Lilly in the pharmaceutical industry and within five years developed collaborations with 35 leading companies in the fields of consumer products, chemicals, specialty chemicals, life sciences, and petrochemicals. These are both US-based multinational companies and European corporations. There is interest in expanding into other sectors as well such as mechanical and electrical engineering, computer science, physics, and optics. The benefits from becoming a Seeker are rather obvious: getting innovative solutions to tough R&D problems, gaining access to a global pool of scientific research, and reducing the investment costs for setting up R&D addressing atypical problems.

Solvers are individual scientists or organised labs. InnoCentive has about 80,000 registered scientists from 180 countries (Springer 2005). People representing an R&D department or a university lab may register as Solvers and submit solutions springing from their collaborative research. To enrich this wide database of expertise, InnoCentive has signed memoranda of understanding with universities and research centres all over the world: India's Council of Scientific and Industrial Research, the National Chemical Lab in Pune, the Indian Institute of Chemical Technology in Hyderabad, the Chinese Academy of Sciences, the National Natural Science Foundation of China, the Moscow State University, and St. Petersburg University.

Solvers who express interest in a particular problem are guided to an online, secure and confidential **Project Room**, which contains the detailed description and requirements of the challenge selected. Specifications are quite simple, but Solvers may ask for additional information and receive guidance.

Prior to this they will have signed a 'Solver Agreement' regulating intellectual property rights among the Seeker company, the Solver scientist, and InnoCentive. The agreement foresees three stages of collaboration: (1) registration, under which InnoCentive grants a non-exclusive license to access and use its services; (2) proposal, under which the Solver elaborates a solution to a challenge submitted by a Seeker who grants an exclusive option to acquire the rights to that solution; and (3) acceptance, when the solution is accepted, and the Solver receives a reward.

To submit a solution, the Solver uses a 'Solution Template' to give structure to the various parts of the proposal (introduction, experimental section, references, supporting data, and conclusion). The solution is reviewed by InnoCentive and then forwarded to the Seeker company that sponsored the Challenge. However, one important term is that nothing in the Solver Agreement can be 'construed as requiring InnoCentive to transmit every Proposal submitted in response to an InnoCentive Challenge to a Seeker'.

Rewards are given to solutions that meet all the specifications and pre-defined criteria. Rewards range from USD 25,000 to 30,000 for theoretical work and papers, or up to USD 100,000 for laboratory verified solutions. The Seeker company that posted a Challenge is obligated to pay the reward, even in the case where has no intention of using the specific solution given to it.

In November 2005, InnoCentive won in 'The Business Processes Award' category at the fourth annual Innovation Summit and Awards event, sponsored by *The Economist* magazine. The award was given in recognition of the efforts made to create the first online forum that enables scientists and science-based companies to cooperate and the vision of collaborative research and innovation that underlies this.

Regional Innovation Poles: multilevel regional innovation networks

Regional Innovation Poles (RINPOLEs) illustrate another approach to collaborative innovation. This is a policy introduced by the Greek Secretariat General for Research and Technology with a regional focus only. Planning for this policy commenced in 2002 with the Ministry of Development that supervises the Secretariat General for Research and Technology under a socialist government, but planning was completed and the policy implemented in 2006 under the conservative government that followed. This policy acknowledges Greece's weaknesses in the field of technological innovation, the difficulties in adapting to the ambitious targets set in the Lisbon Strategy, and instead puts forward the idea of establishing robust, knowledge-intensive regional research, technology and entrepreneurialism clusters in those regions which have a critical mass in this regard. The policy was designed by a team of five advisors to the Ministry of Development, in which I was involved, and was then approved via a process of wide-ranging public dialogue and numerous meetings in the Greek regions.

The **core concept** of this policy is relatively simple. Each Regional Innovation Pole is a complex cluster, a mega-cluster, organised so as to limit unrestrained competition and to develop synergies and multiplier effects in relation to new product development. These principles are as follows:

- Each RINPOLE consists of a small number of research and business clusters. For example, the clusters comprising a pole could be: a team of university laboratories; a team of technology firms; a technology park housing innovative organisations; a research foundation; or a team of technology transfer organisations. While the number of clusters (institutional players) is small, the number of organisations participating in them (firms, research labs, technology transfer organisations) should be substantially large to activate innate, emergent technological collaboration.
- The clusters (technological foundations, teams of firms, technology park, etc.) operate autonomously and each advances its own development plan. It selects targets, and utilises both human resources and infrastructure. It follows internal procedures and decision-making principles.
- In parallel, clusters collaborate with each other on multiple levels. They ensure that the actions they promote are complementary. They ensure political support at local, national and international level. They develop new products and innovative actions in tandem. They generate economies of scale and scope. Individual action plans complement each other.
- Each RINPOLE is a major project for the reference region and is organised and run as an integrated project. The principles involved here include joint planning, monitoring, evaluation, and promotion. They also include horizontal actions that promote synergy and a joint presence on the market.

The adoption of the RINPOLE approach seeks to address the weaknesses in Regional Innovation Strategies (RTP, RIS and RITTS) prepared by many Greek regions in the period 1995–2004, few of which were actually implemented. The key difference between a RINPOLE and a Regional Innovation Strategy is that in the first the system of innovation developed is focused in sectoral terms on a small number of industry sectors, and is focused on core technologies with clear-cut collaboration between research bodies, technology transfer agencies and companies participating in the consortium forming the RINPOLE.

Each RINPOLE utilises multiple technological collaboration mechanisms which are presented in summary form in Table 8.2. These are mechanisms that promote collaboration between research foundations, technology transfer organisations, and businesses or service providers. Each mechanism entails a different type of collaboration depending on the innovation process it seeks to achieve: RTD associations in the case of new product development; networking with technology providers in the case of technology transfer; valorisation of R&D results in the case of spin-off creation; technology platforms in the case of collective technology learning.

The RINPOLE policy was warmly received by the Greek regions, which competed with whatever consultants and political influence they had at their disposal to gain access to the €25 million financing for phase one (2006–2008). Today, five RINPOLEs are in existence in Central Macedonia (ICTs), Western Macedonia (energy), Thessaly (agriculture and biofuels), Western Greece (telecoms, food safety, and protection of the environment), and Crete (broadband services, biotechnology, and medical technology).

The **Thessaloniki Innovation Pole** is one specific example of the programme designed on the basis of these principles. Its actual name is the Central Macedonia Regional Innovation Pole but in effect the organisations participating in it are from the city of Thessaloniki (hereinafter the Pole).

The Pole focuses on one single technological area, that of ICTs. The Pole covers three interrelated industry clusters in NACE 32 (manufacturing of radio, television and communication equipment), NACE 64 (telecommunication services), and NACE 72 (computer related and other similar services). This narrow choice is counter-balanced by the fact that innovative applications in ICTs involve end-users from all the other manufacturing and service sectors. There were three reasons for choosing ICTs as the strategic focus of the Pole.

First, the ICT-related cluster represents an important and dynamic sector of the regional economy, corresponding to approximately 7 per cent of regional GDP and employment. It is among the rising sectors in the region that is gradually replacing older industries such as textiles, clothing, wood, and paper. ICTs, biofood, and health services now form the core of an emerging new regional economy.

Second, ICT-related industries are outward looking activities. ICT expenditure, investment and production shares are rising in the US and in the EU – albeit at different rates across Member States. In the 1990s, several reasons combined to accelerate ICT diffusion and growth. Technological change, coupled with major price reductions led to a surge in the use of digital technologies. With firms ready to exploit the opportunities offered by ICT, the liberalisation of telecommunications and the growth of the Internet economy – allowing for economies of scale and network effects – brought a new eagerness to invest in these technologies. For instance, in the US, business investment in computers and peripheral equipment, measured in real terms, jumped more than four-fold

Table 8.2 Collaboration mechanisms within Regional Innovation Poles

---

I. RTD association in areas of technological priority: Companies, R&D centres, market brokers introducing new products

- Basic research: funding up to 100 per cent
- Applied research: funding up to 50 per cent
- Demonstration: funding up to 35 per cent
- No funding limit

II. Networking SMEs and technology providers. Technology transfer actions

- Networks of SMEs
- Networks of SMEs and technology transfer centres
- SMEs funding up to 40 per cent of eligible costs
- Max. €300,000 per network

III. Support for the creation of new science and technology infrastructure

- Targeted infrastructure funding
- Public organisations: up to 100 per cent of eligible costs
- No funding limit

IV. Valorisation of R&D results. Support for spin-off creation

- Physical persons with know-how
- R&D organisations located in the region
- SMEs
- Max. support €60,000 per spin-off, 60–100 per cent

V. Regional Technology Platforms

- Defining priorities for R&D investment
- Dissemination of research and technology
- Support: up to 100 per cent
- Max. 3 platforms per region
- Max. per platform €150,000

VI. Training

- General training: 60–80 per cent support (SMEs)
- Specific training: up to 50 per cent of eligible costs for SMEs and 40 per cent for the larger companies
- No funding limit

Source: Based on GSRT (2005)

between 1995 and 1999, and a rapid increase was also detectable in the EU, though not at the same pace as in the US. Expenditure in the EU is lower than in the US, although there are certain remarkable exceptions. Sweden and the UK are at the top, with ICT expenditure of about 8 per cent of GDP (1999). Next, in descending order are the Netherlands and Denmark with expenditure near 7 per cent. France, Germany, Italy and Spain are in the low average range for the EU (5.6 per cent in 1999). The most dynamic European countries in terms of ICT expenditure are Greece, Portugal, Ireland, and Finland. They have increased their share of ICT expenditure in GDP and are near the EU average. In the case of Greece and Portugal, the high growth percentages represent serious investment in telecommunication infrastructure, an investment that the majority of the European countries had already made during the first half of the 1990s (European Commission 2001c).

Third, as shown by the Sectoral Innovation Scoreboard 2004 and 2005, the ICT sector (computer and related activities plus electrical and optical equipment) is the most innovative sector in the EU. In parallel, computer services is the most innovative sector in Greece, and exhibits the top performance ratings among all EU Member States. Hence any attempts to innovate will be spawned in extremely fertile ground.

The Pole's strategy is networking and system-building for leveraging the most important weaknesses in new product development: gaps in the innovation performance of enterprises (new products, patents, business research, etc.) and limited production of intellectual property. It is a strategy stemming from the systemic view of innovation and its principal objective is to create a sectoral system of innovation improving the capability of the ICT-related companies for developing and launching new products on the market. In the case of the Pole, the system of innovation is based on cooperation between ICT companies and R&D institutions: research laboratories, technology transfer organisations, liaison offices of universities, business incubators, and technology management consultants. The system operates on two levels: (1) creating an innovation supportive environment; and (2) establishing partnerships and consortia to develop innovative products and services.

As shown in Figure 8.3, ICT companies are at the epicentre of this sectoral system. The Pole is aimed at all ICT companies operating in Thessaloniki and Central Macedonia: local businesses, national champions, branch-plants of multinational companies, spin-offs from research and academic institutions, as well as their close vendors in other sectors of industry and services.

Forward linkages connect ICT companies with customers and markets. New products are channeled towards other ICT companies, companies belonging to other industries, public administration, and the general population. It has been documented that investment in information technology affects output and productivity growth in all sectors through three separable channels (Stiroh 2001; European Commission 2000): (1) Technological progress allows production of improved capital goods at lower prices, thus raising total factor productivity growth in the ICT-producing sector; (2) Increased labour productivity is the

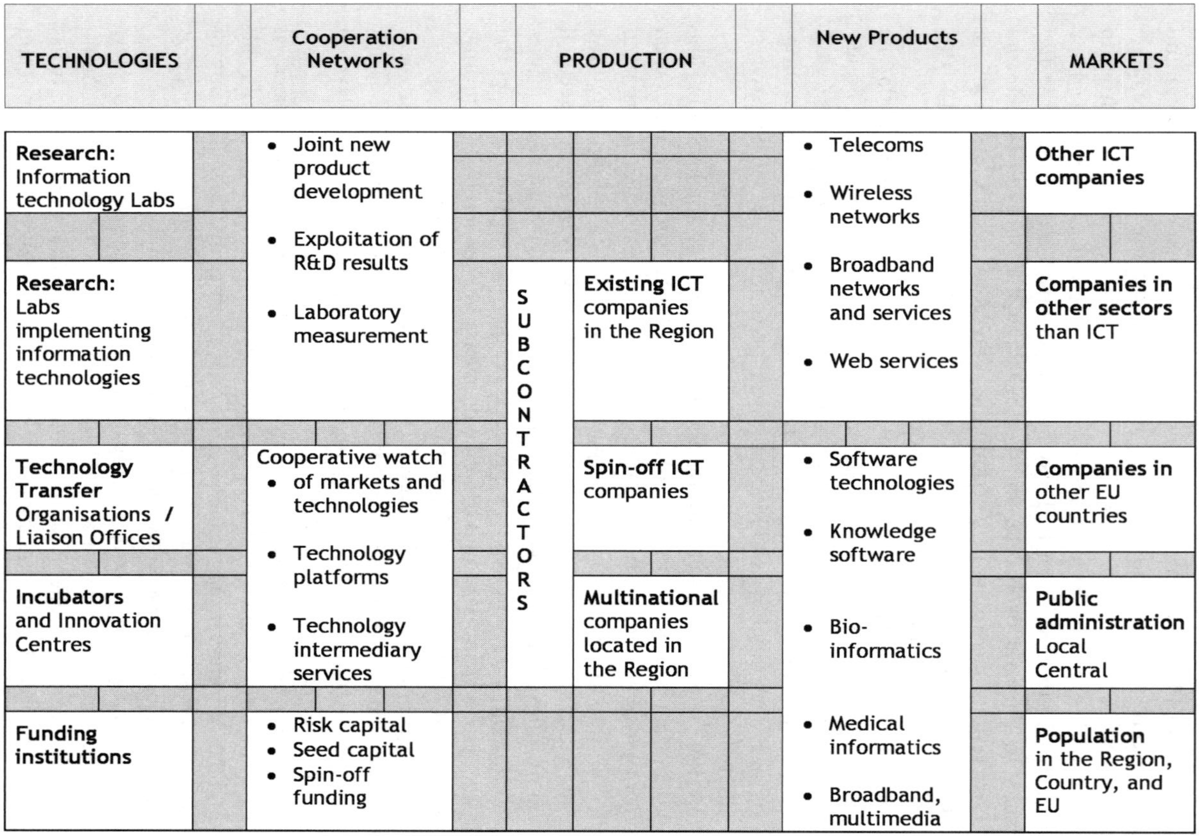

TECHNOLOGIES	Cooperation Networks	PRODUCTION	New Products	MARKETS
Research: Information technology Labs	• Joint new product development • Exploitation of R&D results		• Telecoms • Wireless networks	Other ICT companies
Research: Labs implementing information technologies	• Laboratory measurement	Existing ICT companies in the Region	• Broadband networks and services • Web services	Companies in other sectors than ICT
Technology Transfer Organisations / Liaison Offices	• Cooperative watch of markets and technologies • Technology platforms	Spin-off ICT companies	• Software technologies • Knowledge software	Companies in other EU countries
Incubators and Innovation Centres	• Technology intermediary services	Multinational companies located in the Region	• Bio-informatics	Public administration Local Central
Funding institutions	• Risk capital • Seed capital • Spin-off funding		• Medical informatics • Broadband, multimedia	Population in the Region, Country, and EU

(SUBCONTRACTORS)

Figure 8.3 Innovation system in the ICT sector

most important effect of ICT that spreads out across the total economy; and (3) Spillover effects, as ICT investment induces embodied technological change, thus increasing total factor productivity growth outside the IT sector, generating production spillovers and externalities.

Backward linkages connect ICT companies with R&D, technology transfer, and innovation financing organisations. To achieve innovation in products and technologies companies should engage in targeted collaborative efforts to jointly develop products; use research results; and measure and certify quality, intelligence on markets and technologies, etc.

To make this sectoral system a reality, the Pole has developed four types of actions:

- regional technological platforms in the fields of (1) broadband Internet services, (2) telecommunications, (3) knowledge software;
- new product development consortia between ICT enterprises, research laboratories and institutions, and user enterprises, in the same areas as the technological platforms;
- new spin-off companies based on the exploitation of research results; and
- horizontal activities for the entire ICT sector dealing with the development of the strategic economic intelligence, international technological cooperation, and technology transfer for the creation of innovative entrepreneurship activity.

Regional technological platforms take a cooperative approach when it comes to the selection and application of technologies. A technological platform is set around the agreement of the stakeholders of an industrial sector and their common vision of the technologies that the sector should develop as a matter of priority. Organisations from industry, research, and financial institutions, regulatory authorities, as well as users cooperate to identify both the vision and technologies that can make this happen. Three technological platforms were chosen based on the *Regional Technology Foresight Exercise* that was recently concluded:

- Broadband networks and Internet services: Internet and e-commerce services; distance learning, tele-medicine and mobile office applications; e-governance; business-to-business transactions; digital cities and e-government.
- Digital systems and telecommunication systems: Integrated telecommunication solutions; wireless local networks; hard-wired and wireless satellite networks; systems and communications security; smart buildings and homes.
- Software technologies and knowledge software: Software for industry and commercial businesses and data management; banking, education and training applications; scientific software; physical communication with computer systems, and simulation systems; research management and technology transfer software.

All three platforms operate in parallel using the same methodology. Technologies are defined taking into account the strategic needs of the actors involved and the selection criteria of the technologies in each thematic field ensure a cooperative approach, the transfer of know-how from technology developers to enterprises (end-users), the quantification of the results (e.g. increase in productivity, improvement in the product quality, protection of the environment, saving energy and resources, improvement of health, improvement of competitiveness, creating of new jobs, etc.), and the outline of innovative, promising ICT technologies and products on international markets.

New product development consortia bring together ICT companies, end-user companies, and technology providers from universities and research centres. Each consortium emerges to address the making of an innovative product or service. Following an open call and double assessment, 14 RTD consortia were selected out of 70 proposals submitted, taking into account the strategic importance of the proposed technology application that the consortium intended to develop; a clear evidence of the usefulness and viability of the new product or service; and a clear evidence of consortium partners in a long-term commitment for continuous cooperation and effort to place the new product/service on the market.

The 14 consortia selected have undertaken, with the financial and institutional support of the Pole, to create innovative products and services applicable in cutting edge technology sectors, but also in traditional industries as well (e.g. food, chemicals, metal, plastics etc.). These include ICT-based solutions focusing on digital cities; broadband weather imaging; inter-functionality and adaptability of business-to-business transactions; position-led identification and supply of telematic services; telematics to manage dispatch calls for fleets of vehicles; smart house tele-commands; digital accuracy-driven agriculture; advanced semantics techniques in coronary ultrasound tests; software optimisation for polymer productions; electronic tractability in the dairy industry; quality control based on artificial vision; optimising anti-seismic software for bridges; greenhouse integrated management software; realistic scenery representation in a virtual reality environment; and e-consulting.

The creation of spin-off companies and commercial exploitation of research results is another action of collaborative networking. The birth of new enterprises is a critical path and empowerment strategy for the development of high-tech activities. For the ICT sector especially, spin-offs established to commercially exploit public R&D are a classic pathway for innovation and growth. Efforts in this field are intended to create new knowledge–intensive companies that are based on the utilisation of research results. As in the case of product development consortia, spin-offs represent cooperation between R&D labs and motivated people having the necessary skills to set up and run a new business.

Finally, **horizontal activities** create wider cooperation networks involving all actors related to ICTs. They seek to offer innovation support services to all organisations comprising the ICT innovation system: enterprises, research laboratories, consultants, and technology transfer companies. The rationale behind horizontal activities is open up networking in the fields of business intelligence, market promotion, international cooperation, and technology transfer.

- Business and cluster intelligence are offered to all companies through the systematic monitoring of markets, technologies, and competitors, with a view to improving business management, discovering new markets, and assessing future needs. Business and cluster intelligence are implemented on two levels. At sector level, with the design of an information system addressed to all the ICT companies of the region enabling a continuous monitoring of market and technology trends and company advice; and at the business level, with the dissemination of commercially available applications and software tools for business intelligence.
- International cooperation and product promotion are about the brand name of the Pole, its identity, logo, and marketing campaign. It is a two-edged campaign, on the one hand promoting products and services of the ICT regional cluster, and on the other opening up opportunities for outsourcing in innovation: product concept, engineering, software writing, and accounting. Significant technology clusters have been created in the developing world (China, India) that combine scientific skills and a competitive cost of services. It is now a common belief that in ICT production a competitive presence on markets is impossible without the development of sub-contracting with suppliers from developing countries.
- Technology transfer for innovative business activity deals with inherent risks and uncertainty regarding the final outcome of start-ups and spin-offs. Providing specialised support and evaluation services prior to assuming the business risk is a significant instrument in empowering and accelerating the commercial exploitation of research results. These actions are combined with existing support and finance structures (incubators) for innovative ventures. Parallel to this, collaboration among incubators ensures homogenisation and complementarities between the services provided, creating a strong support cluster for generating new knowledge-intensive businesses.

The Pole is set up as regional non-profit association by 46 organisations: the regional authority, two universities and many labs, one higher technological educational institute, a national research centre, business incubators, a Technology Park, two business associations, numerous ICT companies, specialised ICT providers, and technology dissemination agencies. It is run by a directing board of nine members from the academic, business, and public administration world, and chaired by the General Secretary of the Region. A management team has undertaken daily operation tasks. An independent evaluator monitors the progress made with platform actions, R&D consortia, spin-offs, and horizontal activities, and reports to the association. The actions are implemented by the organisations participating in the association that is the Pole. This wide-ranging participation highlights the interest in innovation through collaboration, but also an awareness that innovation is a top priority in the region. The

New product development based on vertical networks

The third experiment in collaborative innovation that we will discuss concerns networks created along the innovation supply chain. This type of innovation was usually found in industrial districts and vertical clusters organised along a supply chain. Now it is offered as alternative option to the classical solution of innovation within the R&D lab:

> A few years ago a furniture company flew me down to their headquarters to talk to them about innovation, and to get my comments on a new product that they'd developed for the professional services industry. This was a company that had been honored for years as one of America's most innovative companies, so I wasn't sure how much I could help them. They ushered me first into the R&D department where I met with some very creative individuals who obviously knew a lot about their business, and about product innovation. The department featured a giant furniture 'playroom', stocked with a variety of furniture components, where creative minds could serendipitously experiment and build makeshift prototypes on the fly. I was impressed. Being a consultant, the first question I asked them was about their innovation process. Specifically, I asked, how were customer needs, complaints and ideas routed from the front-line customer contacts (the sales and marketing people) to R&D. I got blank stares.
>
> (Pollard 2005)

The classic solution for new product development draws its strength from a well-organised R&D lab, from testing and checking development within the lab, and from direct collaboration within a small team of researchers. This also happens to be its weak point: it is based on collaboration between a limited number of individuals, on the creativity of a small research team, on the limits of technological renewals and any understanding of the market this team may have.

The alternative model is different. It is an open, complex network of innovation experts. It may bring together the skills and creativity of a large number of scientists, engineers, researchers, suppliers, clients and synthesise these skills into a more clear-cut effort to innovate. In this model, the collaborative network can take diverse forms, such as the dynamic of communication and idea sharing. The emphasis may be on the supply side with the participation of specialists and highly creative individuals or on the demand side with the participation of ideal customers and end users.

Comparing the two models, Holmes and Glass (2004) have assessed the factors that have eroded R&D effectiveness in recent years. They argue that many large companies have invested in R&D without reaping significant benefits while being outperformed by competitors with smaller R&D budgets. Why do some companies succeed where others fail? The answer lies in the ability to establish an innovation process involving multiple sources of knowledge and creativity from internal R&D, acquisitions, joint ventures and licensing, and manage this network as a portfolio of opportunities.

The rationale underlying and problems in managing innovation supply chains were explored as part of the New Product Development Networks (NPD-Net) project. This was an INTERREG IIIC project run by a consortium of 6 research bodies: URENIO Research Unit, University of Wales, Research Institute of Human Resources of Panteion University, Institute Josef Stefan, Tartu Science Park, and LEIA Technological Centre.

NPD-Net is about cooperation between specialised technology/innovation providers along the supply chain for new product development. The key concept is 'distributed product development', which highlights the splitting of the product development process into separate tasks (idea generation, screening, design, prototyping, production setting, product promotion, etc.) that are executed by specialised organisations, bound together by institutional agreements, common standards, and continuous communication. New product development is conceived of as a network of actors/organisations, specialised in different stages of new product development, and located in one or more regions.

In NPD-Net this concept led to the creation of vertical new product development networks, each of which had four constituting elements: (I) a NPD Centre with expertise in managing product development in an industry; (II) an online roadmap of tasks, methods and tools assisting new product development; (III) a network of expert organisations with various skills in product development in the same industry; and (IV) a marketing new products service.

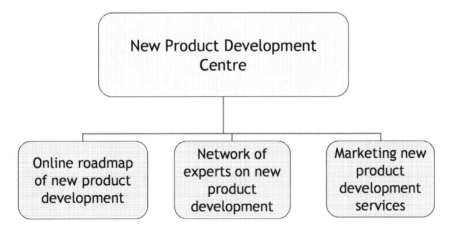

Figure 8.4 NPD–Net concept

I. The NPD Centre is the focal point and coordinating agency for the entire vertical network. New product development is under the authority of this Centre. The Centre constitutes an impartial access point for companies looking for information and expertise on issues related to all aspects of NPD. Its main objective is to simplify the new product development procedure and offer external expertise and skills on various problems found during product development. It is aimed at companies with in-house R&D and product development departments to which it provides assistance in some stages of product development; and to companies without internal R&D capacity that subcontract all the tasks of product development. Furthermore, the Centre has been designed to address the communication gap between NPD suppliers and user-companies, and raise awareness about the procedures and traps involved in new product development.

Various types of institutions work closely with the NPD Centre: university liaison offices, technology transfer centres, R&D labs specialising in different areas of science and technology, consulting companies, and advertising agencies. Through this extended network, the NPD Centre can offer both generic and specialised services at any stage of product development. It can assist with technology information and licensing of intellectual property in the early stages, and with product testing, marketing, and specialised technology consulting in different industry sectors, at later stages.

Three major activities define the operation of the Centre. First, the Centre describes new product development as a sequence of problems to be resolved and tasks to execute, defining the best methods and tools to deal with these problems, and the competent organisations and experts to deal with them. Second, it investigates the demand for new product development, defining markets and target groups, and setting up the experts' networks capable of responding to the demand. Third, the Centre intervenes between supply and demand, raising awareness about network-based new product development and bringing together experts and customer-companies.

II. The online roadmap of tasks and methods is the key organising concept of 'distributed product development'. The roadmap consists of tasks and methods, a suite of actions resolving typical product development problems. Each task may be executed by a different organisation: University lab, research centre, specialised consultant, technology intermediary organisation, consulting company, quality assurance organisation, marketing agency, and so on. Thus, the logical sequence of tasks is transformed into a network of interconnected actors, which cooperate on resolving a defined problem along an established methodology.

A variety of conceptual models have been proposed to describe new product development that differ both in the level of detail/number of activities included in the NPD process, as well as in the level of interactions/feedback that they assume along the different stages and among the different possible players. The process is usually conceived in terms of stages and gates, an understanding mainly due to the work of Cooper and Edgett (2007). While the

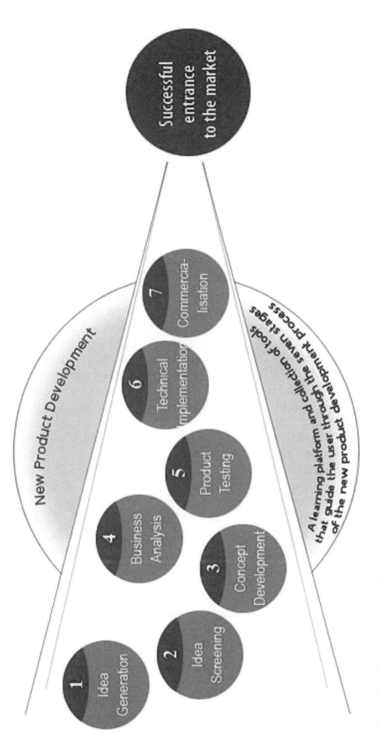

Figure 8.5 Online roadmap of new product development
Source: www.vrc.gr:8080/npd-net/en/npd/index.html

stage-gate model implies a sequential scheme, new NPD methodologies such as Integrated Product Design (IPD) and Concurrent Engineering (CE) follow a parallel/integrated development approach to the different stages with significant interactions and feedback (Fleischer and Liker 1997). This increased interaction has significant implications on the role and management of external sources in the NPD process. In addition, the stage-gate model does not explicitly integrate the use of external sources and collaboration with outside players.

The NDP-Net Roadmap contains a sequence of seven groups of activities called 'levels' and 'control' (assessment) points. Each level contains information and a well-defined series of activities concerned with the particular stage of product development. Each control point is a decision point where senior management can continue or stop the process. To be more precise, a level contains all the information and tools that are needed to successfully complete the particular stage, and the control points contain the required questions or specifications or mandates to which the results of the previous tasks are compared to so that a 'go/kill/hold' decision can be made.

Level 1: Idea generation. This corresponds to a process in which creative thinking is used to produce a large amount of ideas about new products. It is very important that all ideas – no matter how ludicrous or extreme they may sound – be gathered. The idea generation process should be ongoing, have a specific purpose, involve the whole of the company including its clientele, use a variety of methods, have one person in charge and not evaluate the ideas gathered. During the idea generation – gathering process one should not criticise the ideas of others, and one should freewheel and generate as many ideas as possible.

Level 2: Idea screening. Once all viable ideas are gathered, they must be further examined, prioritised and evaluated so that a single product idea is selected for further development. This whole process is called screening and it may be resolved using different tools and techniques. Ideas can be screened against company marketing strategies, against company sales and profitability minimums, along with key customers and buyers, etc. All screening processes should give adequate answers to key questions such as: 'Is it worth it?'; 'Can the product break into the market?'; 'Is it real?'; 'Is it feasible technically speaking?'

Level 3: Concept development and testing. As soon as a single product idea is selected through the process described previously, the product concept has to be further developed. Product concept generation involves defining the target market and customers; identifying the competition and forming a competitive strategy; early preliminary product technical development and testing scheduling; and estimating product development required resources and costs.

Level 4: Business analysis. Once the product concept has been developed, the next step is to check whether the financing is available to back up such a project. Business analysis looks more deeply into the cash flow issue, what costs will arise, what market shares the product could achieve and the expected life of the product. Evaluation of the financial resources needed to develop a concept into a new product can be assisted by 'Cost – Benefit Analysis' in order to calculate expected costs and return on investment; by 'Gannt Charts' to schedule the whole project; by 'Critical Path Analysis and PERT' to manage and relocate resources if necessary; and by 'Stakeholder Analysis' to evaluate support from investors and stakeholders that could influence the development process.

Level 5: Beta and market testing. Product testing can occur at any stage of the NPD process. It can take the form of concept testing at the end of concept development, prototype and beta testing at the end of the prototype development, or final product testing at the end of the technical implementation. Product testing consists of three components: the creation of a testing strategy (which often includes the creation of test cases), the creation of a test plan (which includes test cases and test procedures), and execution of the tests. The test strategy is a formal description of how a product will be tested. The test plan is prepared by reviewing all the functional requirements of the product. The test procedures can define the test conditions, data to be used for testing and the expected results. The test plan should include test cases or scenarios, which should be designed to represent typical and extreme situations that may occur during the product's life. As each test is executed a record of the test results must be kept in a test log. All test results noted in the log must be evaluated by engineers and screened against the pass/fail criteria set out in the test plan. Any faults or bugs in the operation of the product should be fixed before the product goes through the technical implementation and manufacturing phase. When all tests in the test summary are certified, the product receives the go ahead to advance to the next level of development.

Level 6: Technical implementation. Once market and beta testing of the new product prototype having been concluded the new product can go into technical implementation. The product is manufactured in larger quantities so that it can be released onto the selected market or market segment. The problems that are examined concern manufacturing, manufacturing management, production scheduling, and quality.

Level 7: Product commercialisation. In the last stage, commercialisation of a new product, the largest costs to date are incurred. The company will have already spent a lot on product idea generation, product idea selection, product development, prototype development, and product prototype testing and validation, but the amount of money that commercialisation requires is far more important. Commercialisation problems are related to all the marketing aspects of product distribution, product pricing, and product promotion. These aspects include product advertising, product publicity and public relations, product pricing methodologies, selling strategy and selling promotion.

III. The network of expert organisations is created after a systematic survey of new product development supply and demand. Both demand and supply of product development are extremely sensitive to sector. This is particularly true for the more technical aspects of product development, testing, and manufacturing. The NPD Centre and its expert networks have to focus on selected industry sectors to be able to accumulate and offer state-of-the-art services.

In NPD-Net, each Centre specialises in a few industries, which show high demand for product innovation, while there is expertise available for the provision of product development services. In the case of the NPD Centre in Wales, for instance, this appraisal of the current status helped in defining the NPD network and its nodes, taking into account the extent of regional demand, the existence of regional expertise, and the growth potential and sustainability of the NPD Centre.

IV. Marketing NPD services is the fourth element of the distributed NPD. The NPD-Net marketing strategy seeks to resolve two key problems. First, how to manage a classic marketing problem. In other words, how to disseminate the idea of innovation upon a network, to promote the fields of expertise to each NPD Centre, and the individual networks comprised of specialised technology centres that undertake new product development.

On the other hand, there is the issue of how to manage the trust situation. Without doubt, a strong asset of networks established in the context of the NPD-Net is that they accumulate expertise and the quality of services that the innovation networks offer. This is a quality of services that individual new product development consultants cannot offer on their own. Nonetheless, this advantage is also the Achilles' heel of this model. Accumulating expertise in a network entails an increase in the risk for information disclosure. Again we find the 'competence vs. trust' issue being raised. This is the dilemma of each client approaching a NPD centre.

The marketing strategy and dissemination of the collaborative model of NPD-Net does not seek to obscure or downplay the competence–trust antithesis. On the contrary, its objective is to manage the problem of trust and offer solutions to protect intellectual property. The solutions are not intellectual property right registration only but the careful allocation of in-house and external duties related to new product development. In the overall flow of duties presented in the roadmap, the duties retained in-house by firms are strategic in nature. They seek to control progress and ensure links and ties to the overall process.

Collaborative innovation as intelligent space

The three cases of collaborative networks described convey certain messages about how the collaborative innovation model is set up and run and how to achieve the transition from a closed to an open model of innovation. They show that new product development in-house, which the stage-gate models codify and describe, is associated with the wider system of innovation (sector, regional, national, and global) within which it takes place. Successful completion of all stage-gate phases is determined both by the status of the firm developing the new product and by the impact on the wider system of innovation. The three experiments cited show how this link is made via global, local, and sectoral

Partners in NPD Networks in Wales

1 **Idea generation and screening**: BIC Innovation; Light Minds; The Know-How Wales Programme; The Menter a Busnes.

2 **Concept development**: The National Centre for Product Design and Development (PDR) Research; Design Wales.

3 **Business analysis**: Business Eye; Opportunity Wales; SMARTWales; Business Connect; Wales Innovators Network; The Wales Spinout Programme.

4 **Product testing (beta and market testing)**: Business Operation and Management (LERC, Lean Enterprise Research Centre); Electronics and Software (Power Electronics Design Centre, Centre for Advanced Software and Intelligent Systems CASIS, Materials Centre of Excellence for Technology and Industrial Collaboration); Design Engineering and Manufacture (The National Centre for Product Design and Development Research, Manufacturing Engineering Centre, Innovative Manufacturing Research Centre IMRC RCUK Cardiff University, Centre for Electronic Product Engineering); Biotechnology (Aberystwyth BioCenter); Environment (Centre for Research in Energy, Waste and the Environment, The Institute for Sustainability, Energy and Environmental Management, Centre for Advanced and Renewable Materials, CRiBE Centre for Research in the Built Environment, Centre for Complex Fluids Processing); Food (Food Centre Wales, The Dairy Development Centre, Environmental Goods and Services); Magnetics (The Wolfson Centre for Magnetics); Optoelectronics and Bioelectronic (The Centre for Industrial and Commercial Optoelectronics, Institute for Bioelectronic and Molecular Microsystems (IBMM).

5 **Technical implementation**: The Manufacturing Advisory Service Wales (MAS); The National Centre for Product Design and Development Research; Manufacturing Engineering Centre; Innovative Manufacturing Research Centre IMRC RCUK Cardiff University; Centre for Electronic Product Engineering.

6 **Product commercialisation**: The Wales Innovators Network; eCommerce Innovation Centre.

Source: http://npd-net.urenio.org/

networks. These networking spaces are complementary rather than antithetical: regional innovation networks need the assistance of vertical cooperation via production chains that can extend to the global scale transferring skills from all over the world.

Each innovation network generates its own physical, institutional, and virtual space:

- Physical space is determined by the features and geographical location of the organisations which are network nodes: they are research and technology transfer organisations, innovative firms, marketing and market maker organisations. Various combinations of such bodies are possible.
- Institutional space is generated by the agreements and regulatory mechanisms governing collaboration between the organisations (nodes): collaborative research agreements to jointly develop a technology, or exploit or sell technology rights, information non-disclosure agreements or research infrastructure usage agreements.
- The virtual space is demarcated by the communication relations and information management tools which are at the network's disposal: communication networks, innovation management, technology assessment and digital product marketing tools.

During collaboration these three spaces overlap: the physical, institutions and virtual dimension of the collaborative network come together. The collaborative innovation network forms an intelligent space that combines the physical, institutional, and communication aspect of the corresponding system of innovation.

The collaborative innovation model is thus tied into the establishment of an intelligent space with simultaneous physical, institutional and virtual presence at local, sectoral and global level. It is possible to establish this under certain conditions. The network nodes should be innovation generating organisations; creative organisations. They can be located anywhere. However, nodes should be selected based on their complementary knowledge and innovation. Mechanisms regulating collaboration should focus on dealing with problems related to the uninterrupted flow of knowledge with intellectual property barriers being addressed. This is not to say that intellectual property concerns should be ignored. On the contrary, they should be taken into account and regulated via institutional agreements. Institutions should ensure that trust-based relationships are fostered. Virtual spaces should be innovation management spaces. Virtual applications should extend communication, provide information about technologies, markets and best practices and offer knowledge management tools.

From this point of view intelligent spaces may offer a valuable contribution to innovation: they bridge two fundamental contemporary concepts of innovation: 'systems of innovation' and 'open innovation'. Systems of innovation build on the cooperative relationships among the elements that take part in the innovation process, R&D labs, companies, intermediary organisations, funding;

but, imply that these elements concentrate geographically in the national or regional space. Open innovation, on the other hand, praises the cooperation of innovative companies with external sources of knowledge and innovation; the outsourcing of R&D and skills scattered all over the world. Intelligent spaces bring these two spheres together showing how they become complementary. They do it by combining a local nucleus of creative capabilities with institutions of cooperation and digital networks extending over different geographic scales.

9 Digital cities and e-marketplaces

Global promotion of localities, products and services

The last mile of innovation

The promotion and delivery of new products and services is the last mile of innovation when new or updated products created by networks of producers are searching for customers and users. In fact, the concern about product promotion and delivery will have already manifested in much earlier stages of the innovation process. At the start of new product development, estimations about the potential market, market segmentation, customer preferences, consumer behaviour, and other related assumptions heavily influence decisions and choices concerning product values and design. However, the real thing is now, once the new service or product is in hand; when the questions of reaching and persuading customers, companies, consumers, and the public administration to buy are immediate and urgent.

Information technology and digital spaces may play an important role in this last mile of innovation. e-Markets were among the earliest applications of digital technologies to business processes. However, the interest of cities in this field and the creation of digital marketplaces promoting products and services of a city with the direct involvement of citizens, city companies, and the public administration is more recent. This interest goes hand in hand with the rise of Web 2.0, the second generation of web-based communities, in which content is provided by the users through collaboration, sharing, and many-to-many communication. Web 2.0 offers an ingenious solution to a major problem of most digital applications – content – by making the users active participants and content producers.

Community e-marketplaces, e-malls, and online marketing services created within digital cities differ substantially from company portals and privately owned e-stores and e-markets. Their distinctive characteristic is openness. They do not market products and services from a single organisation, but promote those of the city, the cluster, or the community. They create public digital marketplaces in which all interested companies of a community can find a place to 'locate' and offer products and services. They do not operate for the profit of a particular organisation located in digital space but for the interest of the community or city as a whole.

There are also some innovative features in this new form of communication, marketing and promotion:

- Any company, even the smallest one, can have its own e-shop in a digital city marketplace, reach customers on markets, and offer products beyond their traditional reach: the local or regional territory. The cost of maintaining and updating this e-shop is reduced to a minimum, and in some cases may be offered free of charge. Users can add content to the e-shop, in terms of evaluation, recommendation or rejection of products and services, enabling more informed decisions and choices to be made by other customers. Using the tools available, the company can design marketing campaigns and obtain direct knowledge about how customers respond to products and offers. For instance, using Google analytics one can learn about the visits to the marketplace, pageviews, average pageviews per visit, the origin of visitors, what they are looking for more, and so on.
- The public administration can place the promotion and delivery of its own services to the citizen (e-government services) under the umbrella of the same digital space, enabling online delivery of public administration services.
- Promoting the locality is also possible within the same digital space, by communicating local resources, points of interest, natural beauties, and cultural heritage.

The making of digital cities and e-marketplaces, which we will discuss in this chapter, is a concern for the city rather than the individual company that acquires digital space to locate. It is an urban development issue, which requires the most appropriate scenario of network infrastructure and digital services corresponding to the needs of the local population, the local investment capacity, and the development perspective of the city to be identified. It is a telecommunications problem, which requires a sound technical solution and combination of wired and wireless networks in affordable costs to be defined. It is a multimedia problem, requiring the design of digital services in the fields of promotion, marketing, and service delivery open to all companies and citizens. It is a financial engineering problem, requiring a business model of sustainability that respects the public and non-profit character of the digital city to be found.

Digital cities: broadband networks and online services

The term 'digital city' denotes an area that combines broadband communication infrastructure with flexible, service-oriented computing systems. These new digital infrastructures seek to ensure better services for citizens, consumers, and businesses in a specific area. The geographical range of digital cities varies from a small part of a city to highly populous metropolises.

Building blocks of intelligent cities

Digital cities are based on interdisciplinary research in which urban development and governance practices, telecom networks and multimedia applications converge. The term was employed for the first time in De Digitale Stad (DDS) in Amsterdam in Holland as a metaphor for the city's public space. The DDS combined features of a community of people, a website, and a platform for virtual communication between local administration and citizens (Besselaar and Beckers 2005).

Although wired and wireless networking is an important element in setting up a digital city, the local network is only the first step. A digital city requires broadband infrastructure, but is a construct extending far beyond the broadband network. It primarily provides cross-functional services based on the Internet and utilises, where available, existing broadband networks. The network and services contribute to the development of the local economy and improved governance both in relation to in-house service provisions between departments and employees, and in relation to its external relations with citizens and business people.

Today, with a large number of instances in all cities worldwide, the term 'digital city' is used to describe a combination of telecom networks and multimedia applications that support many aspects of the social and economic life of cities: information provision, governance, e-commerce, security, health, education, production, entertainment, transport and others.

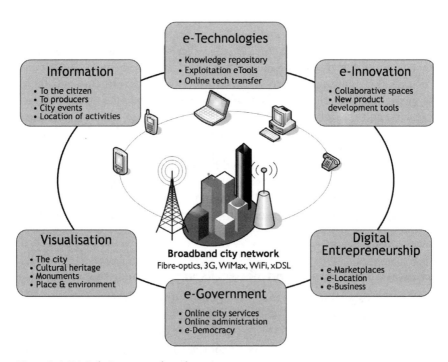

Figure 9.1 Digital city network and services

Were one to examine these applications, one would find that despite their large number they tend to fall into a few large categories such as:

- Commercial digital cities, which focus on providing commercial information and transactions such as the digital cities developed by America Online (AOL) (www.cityguide.aol.com) for the largest cities in the US. These are information portals adopting a 'yellow pages' format, which provide information about local events, hotels, restaurants, commercial stores, cinemas, theatres, and other entertainment options, useful phone numbers, etc. along with advertisements for various products and services.
- Digital governance cities, which are run by municipalities or other public administration bodies offering e-governance services. In their simplest form, they make communication between citizens and municipal authorities easier while in more advanced forms they offer administration services over the Internet (Lévy 1996).
- Virtual cities, which depict the city using either 3D models of buildings or public spaces or 360° panoramas or even videos. They offer virtual tours around the city's public space and the interior of the most important buildings and monuments. To provide these virtual tours there needs to be a broadband network for communication between citizens and the various service providers in the city. One of the first examples of a virtual city was developed by Helsinki which utilised the city's metropolitan broadband network to provide a virtual tour of the city (Linturi and Simula 2005).
- Collaborative digital spaces, which focus on providing services online and enhance the activities taking place in a city. They include distance education, e-learning, technology transfer, health services, home care, and others for example.

These categories of digital cities give a direct response to what digital cities are used for. They provide information about the city, its activities, its people, services, job opportunities and entertainment options; they make commercial transactions easier and reduce transaction costs; they support the online provision of services in education, health and training; they create new opportunities for economic activity, enabling the attraction of enterprises and investments that require broadband infrastructure; they improve the city's image by digitally promoting and recreating it; they improve city governance which becomes more democratic and participatory; they facilitate relations between public administration and citizens; they encourage local democracy and responsibility via greater citizen participation in decision-making.

The most important advantages arise, however, from the combination of physical and digital space within the city. The city's physical space acquires a powerful communication and information processing dimension that increases the functionality of all urban activities. Cities become more accessible and functional due to the new communication capacity and the social dynamic that comes out of it.

The space of digital cities is a representational space, but above all an instrumental space. The distinction between digital and virtual cities refers to the degree to which physical space is represented. Virtual cities recreate and represent the physical space to a more or less accurate degree with varying concepts of representation. On the contrary, digital cities create an intellectual construct for communication correlated to the social space of the city. A digital city may be fully lacking in virtual characteristics: it may not represent but simply accompany the city's physical space. This instrumental, functional dimension is based on communication content. It is a symbolic, non-virtual recreation of the space and functions within the city.

Broadband network

Dedicated local broadband networks are important but not indispensable for a digital city. The alternative is to have the digital city run on the xDSL broadband network of a telecom company. Prices are falling and a community can enter into an agreement with a telecom company and get a good package deal to their citizens instead of developing its own broadband network. However, other options are also available.

Local broadband networks are offering new advantages to cities and regions. As the world goes flat, companies are becoming extremely footloose and they can move parts of their activity to India, China or elsewhere for a fraction of costs in Europe or the US. In the long run, cities and regions in the First World will lose part of their skill and R&D labour base, unless they invent new ways to balance this trend with more creative skills and smart infrastructure able to offer new opportunities and other types of advantages to the low labour cost of developing countries and regions. Local broadband networks are part of such strategies making cities and regions smarter, more efficient, and globally competitive.

Equipped with their own local broadband networks, cities become more attractive, while new job and entrepreneurship opportunities are offered to their residents. The development of a comprehensive, citywide wi-fi network, for instance, is essential to the sustained economic and intellectual growth of a city, its residents and businesses (Kelly 2005). The network and e-services can open previously untapped markets to local companies and organisations, enhancing their access to new customers and revenue. Second, the network can help in bridging the 'digital divide' by offering access to low-income residents who cannot afford the commercial Internet rates of private providers. Third, low cost or free Internet and e-government services are an incentive for increasing the use of digital technologies by the population.

At the start of 2006, for example, the city of St. Cloud, a suburb of Chicago, opened the first totally free wireless broadband metropolitan network in the US. Residents can use the Cyber Spot network free of charge as a public service. In the past the city's 28,000 residents had to pay $600 each a year on average to have broadband access to the Internet at home. One month after the network became operational, 2,972 households and businesses had joined the system while managers estimated that within six months there were 8,000–9,000 registered users from a total of 10,000 households and 1,000 businesses (MuniWireless 2007).

Likewise, the company MetroFi, which operates out of Silicon Valley, offers free wireless broadband Internet access via its own network installed in a series of cities in the area. At present the network covers Santa Clara, Cupertino and Sunnyvale. Connection speeds are up to 1 Mbps when downloading files and up to 256 Kbps when uploading. The cost is defrayed by advertisements (MetroFi 2007).

A digital city's broadband network can incorporate all organisations, infrastructure and people in a city. Technically speaking, it is possible to include every location, function, service, organisation, and physical person in a city in the digital space. As broadband expands, multimedia applications are becoming more interactive and the population of cities are participating in a two-way process of communication and service exchange by downloading and uploading information and digital content. Anyone connecting to the local e-service network enjoys information, governance, commerce, security, education, health, tele-working and other services which have been designed to facilitate citizens. The digital city starts out as a two-way communication phenomenon but extends greatly beyond it.

Establishment of a local broadband network is based on a series of technologies that can be used on a standalone or combined basis. The network can be wired, wireless or a combination of both. Wired networks are based on fibre optics and xDSL technologies. Wireless networks are based in Wi-Fi and WiMAX technologies and on 3G/UMTS mobile telephony technologies.

In the case of fibre optic networks, the broadband access architecture is referred to Fibre-To-The-Home (FTTH) and consists of fibre optic cables terminating at the subscriber's premises. Among the many advantages of fibre optics technology is the high bandwidth, which is many hundreds of times more than that of an ordinary cable; low signal breakdown; low energy requirements; and an absence of free signal transmission waves. Thanks to their small dimensions and durability, fibre optics can be incorporated into existing city networks.

Digital Subscriber List (DSL) is a technology that allows data to be transmitted at high speed via existing phone lines, the vast majority of which service

Table 9.1 Broadband networks and technologies

Wired broadband technologies	Wireless broadband technologies
• Fibre optic networks • Copper xDSL networks	• Wi-Fi • WiMAX • 3G/UMTS • Satellite Internet

Table 9.2 Characteristics of xDSL networks

Type of network	Max. data transmission	Max. data reception	Max. distance
ADSL	800 Kbps	8 Mbps	5,500 m
HDSL	1.54 Mbps	1.54 Mbps	3,650 m
VDSL	16 Mbps	52 Mbps	1,200 m
SDSL	2.3 Mbps	2.3 Mbps	6,700 m
MSDSL	2 Mbps	2 Mbps	8,800 m
RADSL	1 Mbps	7 Mbps	5,500 m
IDSL	144 Kbps	144 Kbps	10,700 m

Source: Based on http://ru6.cti.gr/broadband

a city's telecom needs. The 'x' in xDSL indicates the existence of multiple different and incompatible standards that meet varying needs. This is a technology that has been widely adopted and is the preferred solution for local broadband connections since it uses existing networks both in the city and within buildings.

Fibre optic networks are usually combined with Wi-Fi wireless local networks. The term comes from Wireless Fidelity and denotes a high frequency wireless local network (WLAN). It allows computers in a specific geographical area to interconnect and to connect with the Internet. Data is sent and received using wireless.

Most wireless local networks (WLANs) are based on the IEEE 802.11 standard. Any networks compatible with that standards are known as Wi-Fi networks. Each 802.11 wireless network includes four basic modules: (1) an access point (AP) which is the module that bridges the wired and wireless network; (2) a distribution system which links the various APs in the same network allowing them to exchange frameworks (IEEE 802.11 does not specify how that is to be done); (3) a wireless medium which determines the various physical layers that use either radio frequencies or infrared rays to transmit frameworks between the wireless network stations; and (4) stations which exchange information over the wireless network that are usually either laptops or PDAs, but this is not an exclusive requirement.

In 2003 the IEEE adopted the standard 802.16, also known as WiMAX, to meet the requirements for broadband wireless access (at fixed rates). One important difference between the IEEE 802.16 standard and 802.11 is that the first can be used under conditions of non-visual contact with transmission rates much lower than 50 Mbps. WiMAX was primarily designed to cover Point-to-Multipoint (PTM) connections without precluding its use for Point-to-Point connections.

3G/UMTS networks are wireless mobile telephony networks. The initials stand for Universal Mobile Telecommunications System. They are networks offering satisfactory capacity and transmission speeds. In its initial phase the UMTS network offers data transmission rates up to 384 Kbps where there is increased user mobility. On the contrary, when the user remains stationary the transmission rates increase greatly, reaching up to 2 Mbps. In the near future a further increase in the data transmission rate is expected. 3GPP has already set two new technologies as the standard: High Speed Downlink Packet Access (HSDPA) and High Speed Uplink Packet Access (HSUPA). These technologies are a development of UMTS and hold forth the promise of data transmission rates of up to 14.4 Mbps on the downlink and 5.8 Mbps on the uplink.

Online services

Providers develop and offer digital services over the local broadband network. Various entrepreneurship and viability models allow for differing network and service combinations. In the city of Vasteras, Sweden, for example, a fibre optic network has been developed offering open broadband access to individual homes (FTTH). The term MalarNetCity is used to describe that part of the city interconnected by fibre optics. Network construction commenced in 2000 and the project was undertaken by the local water supply, heating and power utility company, Malarenergi. Today the network encompasses 30,000 homes, 1,800 businesses and schools, public enterprises, medical centres and other organisations. Malarenergi simply provides the network infrastructure. The network is open to service providers who, to date, have developed 85 different services aimed at residential and business customers: these include Internet, television and phone over the web, online games, backup, outsourcing, info services, among others. The network is the largest open access fibre optic network in the world and was dubbed by the Broadband Properties magazine as 'the most advanced Fibre-To-The-Home network' for the year 2006 (MalarNetCity 2007).

Using the local network, a large number of providers may offer digital services. The digital city, like all cities, in not the product of one organisation that plans centrally and implements the entire venture. Just like physical cities, functions are developed on the basis of public infrastructure and services provided by the city's entire population. That is a substantive element of both physical and digital cities. It reflects the historical tradition of cities promoting cooperation and combining resources, infrastructure, and skills. Digital city services relate to the same fields within which physical services are provided: production, exchange, management, and reproduction services.

While the most common digital city services today are still restricted in numerical terms, there is undoubtedly a trend towards increasing the number of services provided. The main categories of services currently available are:

- **Online city promotion**: This offers virtual tours of the city using digital maps and panoramic photographs. At the same time such services can also offer a range of information about culture (such as monuments, sights, events, etc.) that help city residents or visitors organise their free time in the city around their interests.

- **Information**: These are applications providing information about events in selected fields of interest depending on the special features of each city. Information can relate to both residents and businesses operating in the city. With blogs, information is now being provided by everyone to everyone else.
- **Digital entrepreneurship**: Such services provide firms in the online area with tools that support entrepreneurship (business and marketing plan development, market research). They also support e-commerce services offering firms the ability to promote their products via the city's e-marketplace.
- **Digital governance**: These are information and administrative services offered by public administration. They give citizens the chance to acquire certificates issued by local government bodies online over the Internet, to submit requests and applications, and to settle outstanding financial debts to public administration.
- **Digital democracy**: These are applications that improve citizen involvement in decision-making and local government. They allow meetings of the municipal council or committees to be watched online. Citizen involvement in decision-making processes is achieved by them participating in online discussion forums, polls and referendums.

Usually these services are offered via three-tier architecture. Figure 9.2 shows just some of the technologies used in each tier. The first tier is the presentation tier. Users of the digital platform come into contact with this tier. This tier is responsible for the look of the digital city; in other words how information and the services it offers are presented to users. The second tier is the logic tier. It contains the programmes needed to receive user requests from the presentation tier and then make suitable queries in the next. The data from the third level is then processed and it returned to the first. The third level is the data tier. It consists of databases which store both the data from the core and the various modules.

The digital city services include all digital information, entrepreneurship, governance, education, management and other services offered to people and organisations located in the city. The services are much wider than digital governance applications developed by city administration. However, in many cases this distinction escapes us and we consider governance sites to be digital cities. A similar confusion in physical space would be taking the city hall for the city itself.

Of all potential digital city services, let us look in more detail at the product and service marketing section which we have dubbed the last mile of innovation: the digital city marketplaces. e-Marketplaces that are developed within digital cities relate to products and services offered by city businesses, promoting the city itself overall and the services offered by public administration to city citizens. These services are characteristic cases of glocal services, having a distinct global and local aspect. One aspect that relates to marketing and promoting the city, the products manufactured by it, the advantages offered to businesses and

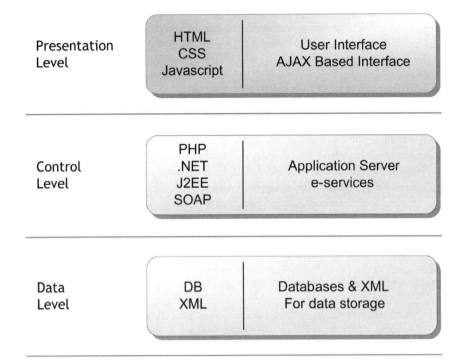

Figure 9.2 Layers of digital city services

talented individuals establishing themselves there, is aimed at the international market and, via the Internet, at a global audience. A second aspect, which relates to governance services, participation in decision-making and local democracy, health and education services, is aimed at the local market, at the city's population and citizens. Their synthesis leads to glocal services as was the case with Digital Birmingham, which under the motto 'going global – looking local' offers services in two discrete sections: 'business' and 'people' (Digital Birmingham 2007).

e-City marketplaces

Digital entrepreneurship covers a huge field of applications. It is continuously expanding due to an ever increasing number of digital customers and digital business support systems. Digital entrepreneurship augments the skills of entrepreneurs in seeing opportunities and translating them into profit-making businesses. It enters into all fields of businesses: markets, processes, skills, and location. In the area of digital cities, it takes a community form as the companies belonging to a locality develop a collective/cooperative presence on the Internet.

Many municipal authorities develop digital commercial centres at city level seeking to bolster entrepreneurship in the area by offering business an additional channel for distributing their products free of charge or at low cost. This helps them channel their products outside the city's geographical boundaries; at the same time, businesses take a step towards the digital economy. A central element of digital entrepreneurship at city level is the establishment of an e-marketplace for all businesses in an area. Businesses are classified into categories and per geographical district of the city, and each entry may be linked to the business' own website. The city's collective digital space offers a meta-space that regulates access to the individual digital spaces of businesses.

The operating rationale of e-marketplaces is simple and is based on setting up an online platform where suppliers and buyers can meet. This platform is used for the purchase and sale of goods and services. Suppliers publish online product, service, and information catalogues. Buyers seek out, compare, negotiate, and select products in real time. All transactions are conducted in a safe, secure environment. Three parties are involved in such digital marketplaces: buyers, suppliers and the operator that has set up the information system and which is responsible for running the e-marketplace:

- The Buyer seeks out opportunities to purchase or procure services and products. The buyer can view and examine product catalogues from various suppliers, add products and services to purchase requisitions and send them to the supplier. Moreover, the buyer can request offers for specific products and services directly from specific suppliers or selected groups of suppliers.
- The Supplier provides the products and services aiming to meet buyer needs. The Supplier ensures that data and information about the products and services in the catalogue is provided and updated constantly. Moreover, responds to purchase requisitions by preparing and dispatching the products requested by Buyers, and at the same time, responds to requests for offers by stating prices for the products or services requested.
- The Operator organises the online marketplace and is not an agent of the customer, does not take part in commercial correspondence and negotiations between the customer and businesses in the system, nor is it involved in the transactions between them on behalf of any party, nor does it get involved in their relationships in any manner.

AOL digital cities were among the fist applications that have implemented this system. America Online (AOL), the largest Internet service provider in the US, has created digital cities for many major US cities that offer access to food/leisure and entertainment/commercial businesses, hotels and property purchase and rental firms. AOL city e-marketplaces are organised around logical categories and the design ensures extensive oversight of all individual business categories. Companies promoted on these e-marketplaces are evaluated, either by special associates or by visitors. The city space is not representative. However,

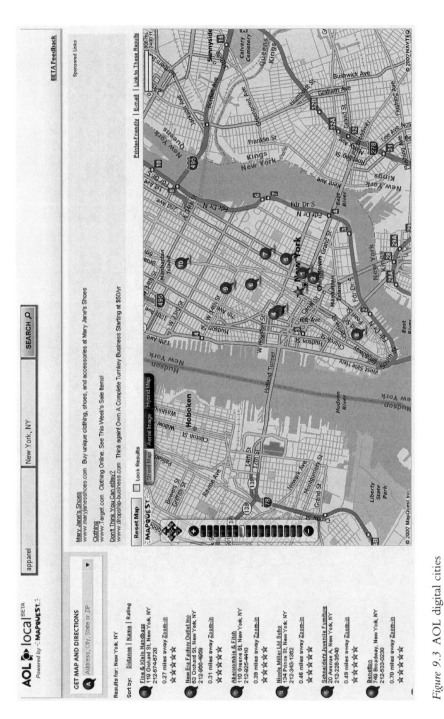

Figure 9.3 AOL digital cities
Source: http://local.aol.com

Figure 9.4 Corfu – digital city and e-market place
Source: www.digitalcorfu.com/vtour.aspx?lang=en&bm=0&id=45

recently AOL in cooperation with MapQuest has developed a Beta version of 'AOL Local', in which firms and stores are shown on a city map and users, by giving their address, get driving directions to go there (Figure 9.3). Stores (movies, restaurants, bars, music, apparel, bookstores, etc.), hospitals, education institutes, colleges, and any other type of organisation are located into a US city sort by name, distance or rating. The locations of firms resulting from a query are plotted on a street map, aerial image, and hybrid map.

In other applications the city's digital space can be presented as a detailed 3D representation of the physical city and the individual website of each business is linked to the city's virtual 3D space. In this case the virtual city takes on the features of an e-marketplace. The user is taken on a tour around a space which depicts the city and meets the businesses in the form they have in physical space in terms of stores, shop windows and products.

The Digital Corfu website takes a step in this direction. It has been financed by the city's regional administration and presents the city's commercial businesses district located into the historical centre.

The representation of the city is based on 400 panoramic shots of areas and routes within the city where the visitor can encounter stores in their actual location within the physical space of the city's historical centre. The 'Shopping' option gives the user the choice of taking a virtual stroll around the shop windows. The user can choose which windows to view depending on the type of business he or she is interested in. The user also has the choice of opting to view all stores on a street or in an entire area. From the list which comes up, users select a store and the screen brings up a graphic representation based on a panoramic photograph of the store and also provides additional information such as the store name and logo, the most important merchandise, the address and contact details.

One application that accompanies many city e-marketplaces is a business search function, usually provided in a business directory format. Both the business directory search options and the amount of information provided about each business vary. Many alternative search methods are offered: alphabetical indexes; categories of business activity; geographical area or maps; keyword searches for business products and services. The business directory is distinct from an e-marketplace since it does not offer online purchasing potential. However, it is aimed at a larger number of businesses, many of which do not participate in the digital marketplace since their products or services cannot be sold online (such as doctors, engineers, various types of workshops, etc.).

The business dimension of digital cities is not limited to e-marketplaces and digital directories. Cities create complex business environments that impact on business running costs, transport costs, and establishment choices. Many cities have already developed digital applications in this sector that are related to selecting a business set-up location. These are services aimed at businesses intending to set up within the city and helps them choose the most appropriate location based on a series of parameters such as establishment costs, subsidisation potential, tax rates, housing zones, related businesses in the same area, proximity to transport, and general points of interest in the area. The way in which this service is provided varies. In the simplest case there are straightforward, static maps that depict the relevant information. In more advanced versions, information is provided on geographical information systems (GIS). The GIS makes it possible to develop dynamic maps based on the choices made by each user. In this way the user can examine various scenarios and then evaluate them. This makes for a more informed choice since choice is now based on a plethora of actual data.

One of the best applications of this kind covers the US capital district (http://app.dcbiz.dc.gov/map/default.shtm). The Washington DC digital city supports a business set up location selector for new businesses in the city. Housing zones, subsidies, business zones, metro stations, police stations, schools and other such data appear on a city map. The user can zoom in on the map and then mark alternative locations and print the map out. The application also offers businesses the option to handle many of the transactions required to set up a new business (such as filing applications, paying fees, etc.) online. Businesses have a wealth of information available to them about the legislation, initiatives taken by municipal authorities to support entrepreneurship, events of business interest and financing opportunities, among other things. Particular importance is also attached to education and training by developing a series of guides which take new entrepreneurs through all stages required to set up a business.

e-City promotion

This is equivalent to e-marketplaces but for the entire city. Virtual tour applications promote cities by creating digital representations that enable us to get to know the city via its digital image. Such approaches are part of the promotion, marketing and attraction strategies deployed by cities to draw in visitors, consumers and possibly investors. They are addressed to citizens worldwide: potential visitors and consumers who along with discovering the city via its digital image could become potential buyers of local products and services. Especially for cities relying on tourism and well-known cities worldwide, promotion via virtual city tours has become a strong marketing tool.

Virtual tours are organised on at least two levels, which link the overall representation of the city with individual points of interest and tours. The first level depicts the city overall and ensures overall oversight and orientation. This can be achieved in many, various ways: with a list of choices such as sights, museums, entertainment, accommodation and so on; using an interactive map of the city that shows a series of points of interest, offering a more representational image of the city; and using a 3D model which shows the built up and open spaces within the city in a virtual environment and makes navigation easier. The format of the map used as a digital background on which information is displayed varies. In early applications relatively simple maps were used. Following that, more detailed maps were used either with artistic representations or actual models of the city based on satellite and aerial photographs.

In 2005 Google presented the Google Map application (local.google.com), which provided free access to road and satellite maps of cities in the US and Canada (geographical coverage later extended to Europe, Japan, Australia and elsewhere). With the Google Map API anyone can use Google Maps and free cartographic data in their own web applications. This API's open architecture resulted in tens of instances of use by independent users appearing within a short period of time after release of the application. Yahoo (local.yahoo.com) and Microsoft (local.live.com) have followed suit and offered similar functionality.

Likewise, many cities have used web cameras that transmit live images of the city in real time. In terms of city promotion what was shown has not been remarkable: the quality of image is rather poor and static.

The second level focuses on individual points of interest. By selecting a point of interest from a list or map of the city, a detailed presentation or overview is provided. In many cases the points of interests are grouped along specific virtual routes shown on the map. These routes are pre-set or can be dynamically generated by users. Presentations are made using mainly panoramic shots or video and accompanied by text, sound and hyperlinks.

One particularly noteworthy application, which utilises this architecture, is Virtual Canberra. It depicts the Australian capital and stands out for its user interface, its esthetics and the 3D model of the city. It uses an innovative navigation method that allows users to select, search and view urban routes within a virtual environment. The aim of this digital representation is to promote the Australian capital; to improve its image. Although Canberra is the national capital, it is less accessible than other Australian cities like Sydney and Melbourne. Given the major distances separating Canberra from other urban centres in the country, the capital is not particularly well-known, even among Australians. The website seeks to redress the capital's limited recognition by offering a virtual tour of the city's space and monuments.

The first level takes the form of a 3D model of the city. The user can rotate the map in all directions and view the city from different perspectives. This model covers an area of 3,400 hectares and is the largest city model in Australia. The 3D map presents points of interest as blue spheres that the virtual visitor can move to.

The second level focuses on individual points of interest. When a point of interest is selected it changes colour to orange, its name appears on the map and in the upper section of the screen a panoramic shot of it appears. By moving the cursor over the panoramic shot, it rolls in the same direction as the cursor. When pointed in the direction of a neighboring point of interest, a panoramic shot of it appears with a blue arrow and the name of that point of interest. By clicking on it, the user is taken to the new point of interest and the 3D map is also updated. The Showcase option, which is located below the panoramic shot window, provides information about the point of interest being viewed. This information contains text, photographs that can be scaled and a close-up snapshot of the 3D map showing the specific sight. Overall, the city is captured

242 *Building blocks of intelligent cities*

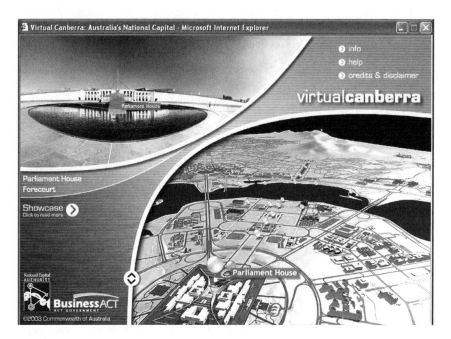

Figure 9.5 Virtual Canberra
Source: www.virtualcanberra.gov.au

in 21 high quality and esthetic panoramic shots. This website was funded by the National Capital Authority and is maintained by a team of in-house developers. A new design of Virtual Canberra was launched on 22 May 2006, and the high quality application described has been removed.

Compared to instrumental digital spaces on the Internet, virtual city spaces are representational. Eventually, they may simplify and limit the complexity of the city's physical space to better match the expectations and requirements of the virtual tour. The time devoted by each user is quite limited and users are just one click away from another digital city.

Virtual cities usually paint physical space in a good light, choosing to present the more positive and interesting aspects, leaving the rundown areas and outdated infrastructure which are to be found in every city out of the field of vision, in digital Lethe.

e-Government

e-Government is the equivalent of e-marketplaces in the field of administration services. It is defined as the use of information and communication technologies to improve the relationships between public administration and citizens, at any level of central or local administration; in particular, the use of Internet-based services provided by the administration to citizens, businesses, and other organisations located in a given geographical area. As described in Webopedia, e-gov

> refers to any government functions or processes that are carried out in *digital* form over the *Internet*. Local, state and federal governments essentially set up central websites from which the public (both private citizens and *businesses*) can find public information, download government forms and contact government representatives ... e-government also refers to the standard processes that different government agencies use in order to communicate with each other and streamline processes
>
> (Webopedia-e-government 2007).

The information system supporting e-government is a centralised one, in which public administration and the users interact. Communication, information flows, and service delivery between the administration and citizens follow a two-way interaction path:

From the citizen to the administration:

- Demand for information or service
- Notification of an event
- Declaration of a state
- Payment of a bill
- Declaration of an opinion, choice, agreement or disagreement

From the administration to the citizen:

- Provision of information
- Provision of certificate or other official document
- Actualisation for the provision of a service
- Actualisation of a decision-making process

Two major areas of e-government are 'e-administration' and 'e-democracy'. Any good e-government system should cover both areas. Robert Bell from the ICF highlights Virginia Beach as case of best-in-class e-government (Bell 2006). The city provides a series of e-government services, including media streaming of City Council meetings, document archives for land use and economic development initiatives, an online forum of public voices on redevelopment projects, e-mail alerts to citizens who opt in on various topics, online service request (mosquito spraying, trash pickup, water and sanitation service), electronic payments to the city, and emergency preparedness including custom emergency (hurricane) reports.

e-Administration can include almost every service offered by a public administration. A precondition is that the way in which services are provided is reorganised so that they can be offered via online communication. An impressive e-administration system has been developed by the city of Barcelona. It allows every citizen or business to submit applications and proposals related to the city and municipal administration. It offers more than 20 access channels with the aim of maximising the number of citizen transactions with municipal services. The obligation is not only to solve any issue referred to and to take into account all proposals received, but also to respond to all requests so that the responsible role of local administration is recognised. In 2004 the system received more than 4 million phone calls and there were 15 million hits on the municipal council's website. More than 800 people are employed by the service (250 operators and 550 people charged with addressing queries). Among the services provided online are: building permits, capital gains tax, tax and fine complaints and appeals, permits for opening premises and setting up a business, direct bank payments to the council, registration or change of residence, certificates of residence, certificates of co-habitation, and tax payment certificates. All citizen requests are treated as an opportunity to improve the quality of life and to demonstrate municipal responsibility since every citizen can monitor the progress of his request.

e-Democracy is also a major challenge for e-government. It can be applied both at national and at local level. The technology to perform electronic elections and other forms of participation in decision-making is available. In fact, some organisations, companies, universities, municipalities, and states already use e-voting to choose their officials. e-Democracy integrates and includes concepts that are still new and constantly developing with many pilot applications to promote understanding of how the use of ICTs can change relationships between citizens and government. There is no model specifying all functions of e-democracy and how electronic voting participation in decision-making complete other forms of democracy (representative, participative and direct).

Electronic petitions (e-petitions) is a form of online democracy enabling wider citizen participation in decision-making. One of the few applications in this regard has been developed by the Scottish Parliament and within the first 12 months of operation it had attracted 20,812 signatures, 639 comments, and 45 e-petitions. The application allows a petition to be posted live on the Internet. The petition can thus be made available to a much wider audience and gather more citizens to support it. Only the name and place of each participant appear on the website. The other details the user gives are needed to validate his/her signature. Each e-petition also has its own discussion forum, in which participants can express their opinion and discuss matters (The Scottish Parliament 2007).

Electronic voting (e-voting) is a more advanced step of e-democracy. Voters simply point and click on the candidate or option they select. This type of voting has the potential to significantly increase voter turnout. The main challenge, however, is security and authentication. To a large extent, security issues depend on the organisation of the process prior to an election. This may involve procedures for registration, distribution of identification passwords, and validation of votes. A typical election demands robust procedures for registration, validation, collection, and tallying. However, doubts about the accuracy and integrity of e-voting equipment have been growing.

On a local level, Geneva is one of the few cities that, in 2003, introduced and uses e-voting on a regular basis. Swiss citizens vote four to five times a year, sometimes more, and this 'direct democracy' is suited to Internet voting, not only because it implies numerous ballots, but also for the many competences resting with the citizens and the limited delegation of sovereignty given to representatives. Electronic voting is not replacing physical voting, postal voting and polling stations, but is offering a third possibility to the citizens (Republique et Canton de Geneve 2007).

Problems with and doubts about the unimpeachability of elections has always existed and will continue to exist no matter how voting is organised. The major contribution of e-democracy is that it allows for direct democracy on all issues in public life and not just the election of representatives. This prospect, and the consequences on the political system, are the reason for the delay in implementing it, rather than concerns about the unimpeachability of the system.

Sustainability of digital cities and e-marketplaces

Are e-marketplaces and the online services of digital cities sustainable? Do they require constant public support and financing? As is the case with most things nowadays, the viability of digital cities and e-marketplaces depends on the business model adopted.

A series of alternative options are available to public/municipal authorities to establish and run a broadband network. For example:

- Public network/private use. The broadband network operates as an extension of the already existing public telecom network that meets the area's needs. The broadband network is run by a separate private company.
- Public network/mixed use. The municipality uses part of the telecom network for its own needs and at the same time operates as an Internet service provider (ISP). The network is either installed by the municipality or a private company.
- Joint operation. The network belongs to the municipality and is run by a joint company which exploits part of it for commercial purposes.
- Private network/private use. The company that the network belongs to offers part of it to the municipal under preferential terms and conditions (free or low cost) to interconnect its services. The rest of the network is used by Internet service providers.

The services provided to citizens and businesses range from simple Internet access (e-mail account, web hosting, computer skills and Internet technology training) right up to the range of services outlined in the foregoing section. Here the business models for providing them are numerous. Internet access services are offered either free or at low cost (compared to the commercial price). In the case of free services, there is an option to provide more advanced services for a fee (such as more bandwidth, security, etc.). Since the cost of putting broadband networks in place is high, the number of totally free services is minimal. In most cases the cost is met by advertising that appears when connected to the Internet. In the simplest scenario these advertisements come from sponsors, but more complex targeted advertising schemes have already begun to be implemented. Targeting relates to the geographical area of the individuals connected to the Internet, to the financial status of users and the type of information viewed by them. The collection and processing of data required to provide targeted advertising raises a series of issues about protecting the user's personal rights and liberties. This debate is underway and methods are being sought that combine the viability of these services with better user protection.

In several cities, free Internet access is available in each home or at each business. The spread in wireless network usage is contributing to the use of free Internet services.

At the end, there is no universal model of sustainability. The solution to the viability issues is tied into the specific features of each city, and the ways in which public and private initiatives can be combined.

10 Building blocks of intelligent cities

Architecture of layers and functions

Two movements shaping intelligent cities

Intelligent cities, communities and clusters are at the intersection of two major contemporary movements: (1) the rising innovation economy and innovation-led development of cities and regions; and (2) the expanding use of the Internet, broadband networks, and e-services that feed the increasing digitalisation of contemporary urban life. Intelligent cities bridge these major trends of our time and improve the placement of human communities that realise this association within the redistribution of wealth and power the same trends bring. Applying ICTs, e-tools, and e-services, the system of innovation within a territory is enhanced in terms of networking, reach, and efficiency. The city gains innovation capability, which is then translated into increased competitiveness, better environment, more jobs and wealth. Out of the meeting of innovation and digital space the two fundamental dimensions of intelligent cities are defined:

- On the one hand, the system of innovation (local, regional, sectoral, global) within which a continuous mix of skills and learning institutions takes place, driving the development of new products and technologies in the organisations located into a territory (companies, R&D centres, intermediaries, incubators, etc.).
- On the other hand, the digital reconstruction of the city and the collaborative digital spaces of knowledge management and innovation. These applications facilitate communication, data storage and retrieval, knowledge transfer, cooperative product development, and product promotion, enhancing localised innovation capabilities.

Intelligent cities evolve along with the trends characterising the two aforementioned dimensions. For instance, global innovation networks and open procedures actually predominate within systems of innovation; and within digital spaces the Web 2.0 and the active participation of users in content development predominate. No doubt, different trends will appear in the future, influencing respectively the content of intelligent cities.

With respect to these dimensions, two planning paradigms also compete in the making of intelligent cities: cyber-cities vs. intelligent communities. Their major difference is on the different weightings they attach to the two aforementioned components of intelligent cities (innovation system and collaborative digital spaces). Cyber-cities consider that the main challenge for intelligent cities lies in the level of the digital networking, human-machine communication, sensors, intelligent agents, and other technologies for automation in information collection, processing, and dissemination embedded into the city infrastructures. Intelligent communities, on the contrary, consider intelligent cities to be a combination of human innovativeness, collective, and digital intelligence; and the challenges of their making are at the integration of innovation capabilities, institutional cooperation for innovation, and digital spaces facilitating this cooperation.

Reviewing the literature on intelligent cities, Radovanovic (2003) further opened the field by drawing a picture that places them at the centre of four blocks of influence: (1) knowledge economy and information society, fuelled by creativity and innovation; (2) intellectual capital; (3) economic intelligence; and (4) future society, and the dream society in particular. The knowledge economy and information society are framework conditions; well-organised intelligence and intellectual capital are the major resources of creativity and innovation in the knowledge economy; the dream society refers to a forthcoming future, which is not made of certainties but of dreams concerning technological breakthroughs and great achievements.

Intelligent cities and regions as systems of collective information and knowledge management institutions existed in the past. Radovanovic (2003) refers to Ragusa, a tiny city-state on the east cost of the Adriatic Sea, which had an impact far beyond its size and power. The city sustained its independence over five centuries on the basis of well-organised intelligence, using all human and technical resources available for collecting and analysing strategic information. Information was one of the central factors that helped Ragusa to maintain a balance between the great powers of its time; between Venice and the Ottoman and Habsburg Empires.

Mazower (2004) attributes the same characteristic to the city of Thessalonica, Greece in the sixteenth century, where commerce and intelligence blended. With the arrival of Sephardim Jews expelled from Spain in the 1490s, the city rose to become a global trading centre linking northwards with the Balkan markets, south and east with the Asian trading routes towards Persia, Yemen, and India, and westwards with Venice and other Italian ports. Italian, Arab, and Armenian merchants all participated in this intense exchange, which combined trade with the best intelligence networks in the entire south of Europe and the Mediterranean region.

Compared to the past, modern intelligent cities and regions have one additional dimension: digital networks, digital communication spaces and web-based intelligence. Among the driving forces of innovation we must now include the strength of intelligent machines and new combinations of individual, collective, and artificial intelligence. This approach is primarily being promoted today by the Intelligent Community Forum which gives prizes for the efforts made by local and regional administrations of cities and regions worldwide to promote the innovation economy while also combining it with the information society (www.intelligentcommunity.org).

Intelligent cities bring closer the most important processes of our times, generating multiplier effects both in accelerating the innovation economy and in deepening the information society. The key questions we need to answer, however, relate to how innovation and the information society are interconnected at the local level, and to the architecture of intelligent cities, defined by the superimposition of innovation systems and digital spaces to support and augment them.

Intelligent cities as territorial systems of innovation

The first dimension of intelligent cities is associated to constellations of organisations forming networks, clusters, districts, poles, and systems, within which innovation takes place. Today, it is a mainstream conception that innovation is systemic. Theories of innovation have radically changed over the past few years. Both the traditional Schumpeterian model, regarding innovation as an internal activity of the firm (Schumpeter 1934), and the linear innovation model in which new product development follows a step-by-step sequence from discovery, idea generation, business case analysis, to product development, testing, and launch have been found to be inadequate in conceptualising this process (Cooper 1994; 1999). Increasingly innovation is regarded as a collaborative and evolutionary process taking place within environments augmenting discovery and idea generation, and selection of the most plausible innovations.

Systemic theory of innovation was initially formulated at national level. Foundational publications by Lundvall (1992) and Nelson (1993) focused on and described national systems of innovation. Gradually however, there was a shift towards the regional and local levels. A series of publications has shown that innovation processes are embedded in regional conditions shaping regional systems of innovation (Braczyk *et al.* 1997; Cooke and Morgan 1998; Simmie 1997). Kaufmann and Todtling (2000) identified five major mechanisms that explain the regional embeddedness of innovation:

- Many of the preconditions of innovation, such as qualifications of the labour force, education, research institutions, knowledge externalities and spillovers, are immobile giving some regions advantages over others.
- Industrial clusters are localised giving rise to specific innovation patterns within networks and industry sectors.
- A common technical culture may develop through collective learning taking place in a regional productive system.
- University-industry links and knowledge spillovers are region specific.
- Regional policy plays an active role in innovation providing support through institutions and agencies.

Innovative agglomerations and territorial systems of innovation (technology districts, technopoles, innovative clusters, technology parks, innovating cities and regions) can be described in terms of (1) constituting institutions; (2) cooperation networks; (3) rules of operation; and (4) innovation outcomes. Their key components come from the business, R&D, technology transfer, and funding sectors: innovative firms; supplier firms; customer firms; universities; research organisations; technology transfer institutions; IPR lawyers; consultants; training institutions; incubators; funding organisations; government agencies; and monitoring organisations. These elements are organised into networks and innovation springs from their synergy. What gives value to components is cooperation. Networks facilitate and augment innovation capability at company level; the latter being the ultimate producer and beneficiary of innovation. However, the connective substance of all networks is knowledge. What flows within innovation networks is, above all, knowledge.

Cooperation networks rely on knowledge flows and on institutional regulation. Institutions for knowledge dissemination, intellectual property management, assessment, and funding, act like gatekeepers or switches which turn funding on and off, and take 'kill' or 'go' decisions along the innovation process. To do so, institutions regulating these flows are placed within knowledge networks linking each organisation with its external partners. All kinds of knowledge flow within innovation networks: declarative knowledge about facts; procedural knowledge dealing with know-how; and conditional knowledge linking conditions and effects (Dawes 2003); 'explicit' knowledge which is transmittable in formal language, codified and captured in libraries, archives and databases; and 'tacit' knowledge which has a personal quality that makes it hard to formalise and transmit in ways other than personal communication.

Knowledge network architecture changes with respect to the innovation process that takes place. Innovation processes such as cooperative R&D, strategic intelligence, product innovation, process innovation, spin-off creation, and opening up new markets, for instance, involve fundamentally different knowledge networks. This should be expected as different processes of innovation engage different partners and forms of cooperation. A cooperative R&D project, for instance, may have a network architecture that differs substantially from a cooperative network for strategic intelligence. The entire landscape of networking is becoming extremely complex and variable. However, the connectivity of components within innovation systems is characterised by two principles: (1) the creation of knowledge constellations and clusters with various internal architectures; and (2) the functioning of knowledge switches regulating the flow of knowledge between the members of an innovation network.

This architecture of **knowledge networks and institutional switches** so characteristic of spatial innovation systems depends on the type of innovation: in the three basic types of innovation (product, process, and organisational) radically different types of knowledge network architectures correspond.

Knowledge networks in new product development

The architecture of knowledge networks and institutional switches characteristic of product innovation is defined by the internal logic and stages of new product development (Figure 10.1).

The core of the network is the new product development process, which can be depicted in a stage-gate diagram (Cooper 1994; 1996; 1999). Stage-Gate systems have been presented with many different names such as PDP (Product Delivery Process), NPP (New Product Process), Gating System, and Product Launch System. The term 'stage-gate' is characteristic, precisely indicating what the system is: a process comprised of stages and gates. In the stage-gate system, the innovation process is divided into various stages and each stage is evaluated by a gate or checkpoint. The project leader collaborates with a cross-functional group and collectively they assess the project in each stage at each gate before the project moves on to the next stage. At each gate one of four possible decisions must be taken: continuation to the next stage (go), rejection (kill), hold, or return to the same stage (recycle).

Modern-day stage-gate processes have their roots in previous models. The well-known stage-gate systems (those which were widely-known in the 1990s) are second generation models. The first generation outline for product development was elaborated by NASA in the mid-1960s (Phased Project Planning or PPP) and was a complex, detailed scheme for ensuring better collaboration between contractors and suppliers. Today it is frequently known as the Phased Review Process. This method divided development up into discrete phases. There were review points at the end of each such phase and financing for the next phase was based on the evaluation of specific parameters that had to be met. Formally, that meant that all pre-specified work had to have been satisfactorily completed in the previous phase. Consequently, this method was more a measurement and control methodology. It ensured that the project evolved in line with initial planning. However, review points rarely addressed project completion from a business perspective. In second generation stage-gate systems, the stage-gates became cross-functional. A cross-functional project team that incorporated diverse areas of specialisation undertook to reduce the impacts and remove the barriers generated by knowledge fragmentation in functional areas. The marketing and manufacturing sectors became integral parts of the product development process. Gates also became cross-functional. Precisely designated decision points with go/kill criteria were another improvement. Today stage-gate systems present strict gates with precise criteria and measurements – criteria that focus on quantitative metrics and qualitative business feature measurements, such as product advantages, synergies and market interest. Third generation systems are already being developed, with particular emphasis being placed on the efficiency and allocation (networking) of development factors and resources. Third generation processes represent an uncertain equilibrium between the need for complete actioning and full information and the need for rapid moves to be made. The process is based on four fundamental Fs:

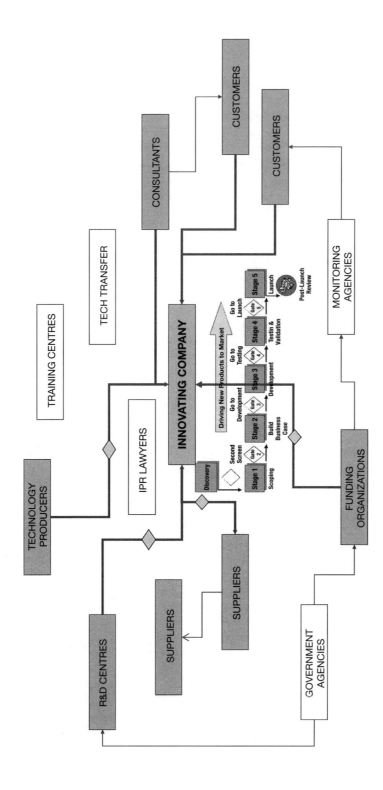

Figure 10.1 Knowledge networks in product innovation
Partners (in rectangular) and switches (rhombi)

(1) *Fluidity*: the process is fluid and adaptable, with fluid stages which in many cases can be omitted to speed up implementation; (2) *Fuzzy Gates* which are interdependent/networked (instead of being absolute); (3) *Focused processes* based on prioritisation methods that are oriented to the overall project portfolio; and (4) *Flexible processes*, which are not rigid and strict since each project is unique and has its own course to follow. The system is more 'intelligent' and adapted to the special needs of each project (Kitsios 2005).

Around the stage-gate process, a wide knowledge network develops that encloses the entire innovation system from research to placement of the new product on the market. In this network feeding new product development, knowledge flows are two-way: from research bodies, technology production organisations and the market to the firm developing the new product as a transfer of technologies and business skills; contrariwise, from the firm to suppliers, as the transfer of technical specifications and technologies for producing new product components.

There are four types of switches controlling the flow of information: (1) institutional cooperation agreements with research centres via joint research programmes; (2) purchase of intellectual property from producers/holders of technological knowledge and IPR; (3) resource flows from financing bodies; and (4) cooperation agreements with suppliers for information and knowledge disclosure.

Knowledge networks in process innovation

Knowledge networks and institutional switches that are characteristic of process innovations are significantly different from those cited above. To a large degree, they are defined by knowledge and know-how transfer relationships, the acquisition of user rights and the ability to absorb selected technologies (Figure 10.2). In this case, the aim of the knowledge network is to acquire technologies which are thought to be important for a particular business. These could include more efficient energy management, automation technologies, broadband installation and management, processing technologies, waste treatment and recycling technologies, etc. Knowledge networks and cooperation agreements are major strategies to acquire state-of-the-art technologies, not available at arms-length relationships. We should underline that technology transfer plays a primary role in innovation here since most of the technologies used by an organisation are not generated by the organisation itself.

Knowledge networks in technology transfer vary according to available resources and the motivations of recipient organisations. Different architectures are set around licensing, cooperative R&D, and spin-offs agreements.

Licensing agreements concern the transfer of intellectual property rights in order to make, use, and sell a certain product, design, or service by a party that has the right to give this permission (Rogers *et al.* 2001). Royalties are the fees paid for acquiring a license. Licensing agreements usually link universities and other technology producers to companies wishing to use technologies. The increasing exploitation of university and public R&D through licensing is offering additional funds to universities and public R&D centres.

Cooperative R&D or *contract R&D* agreements are comprehensive legal agreements to share research personnel, equipment, and intellectual property for a common research objective/project. The network usually links one or more university research laboratories and one or more business, creating a research consortium for a limited period of time, necessary to carry on the research objective.

Spin-off creation offers a mechanism to commercialise technologies that originated from a university lab, a government R&D centre or a private R&D organisation. It involves the creation of a new company by the parent organisation, which undertakes to commercially exploit a technology. Usually spin-offs are formed by individuals who were former employees of the parent organisation. The university, R&D centre or parent company holds part of the company's shares in exchange for the know-how that it transfers to the new company.

More flexible forms of technology transfer through networking also include consultancy and technical services provision, personnel exchange programmes, and training (Lee and Win 2004).

As shown in Figure 10.2, most networking takes place on the supply side, linking an innovative company with technology and R&D providers. The role of customers and suppliers is less important because most process innovation needs relate to rationalisation objectives, cost and waste cuts, and originate internally within the company. In all cases within the network flows codified knowledge; we are dealing with codified and supply-side knowledge architecture.

Knowledge networks in organisational innovation

Organisational innovations relate to the entire supply chain and are associated with better, more effective and more cost-efficient operation of that chain. Organisational innovation in the supply chain seeks to cover a wide range of objectives such as ensuring a steady supply of raw materials and components, fast response to changing needs, minimised transport and distribution costs, minimised inventories, deployment of buffer stocks of parts or finished goods, shortened lead time, and increased quality of partners and collaboration. Well-known organisational innovations that were developed recently include just-in-time delivery systems, lean production, flexible supply chains, and vertical quality certification.

Supply chains incorporate and integrate many different organisations: raw material producers, suppliers, manufacturers, assemblers, warehouses, distributors/wholesalers, retailers, customers, and end customers. Major stages are the production of raw materials, processing, assembly into end products, distribution and purchase by end consumers (Lambert and Cooper 2000). The supply chain can be considered to be a system that consists of logistic chains and coordination activities. The system is made up of two groups of entities: the supply chain partners and supply chain management (Feldmann and Muller 2003). Every partner is connected with other partners. There is permanent exchange between

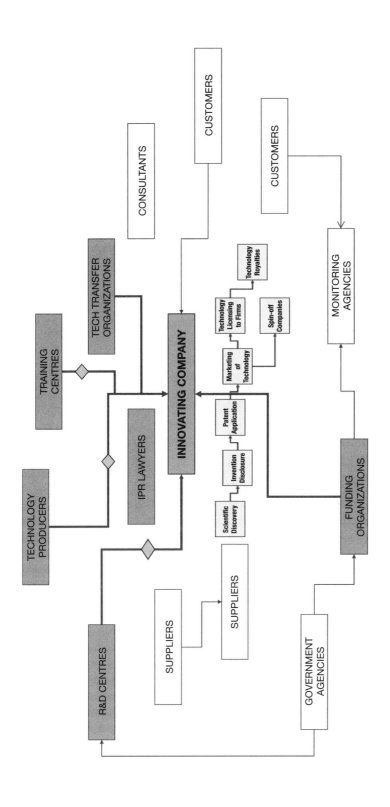

Figure 10.2 Knowledge networks in technology transfer processes
Partners (in rectangular) and switches (rhombi)

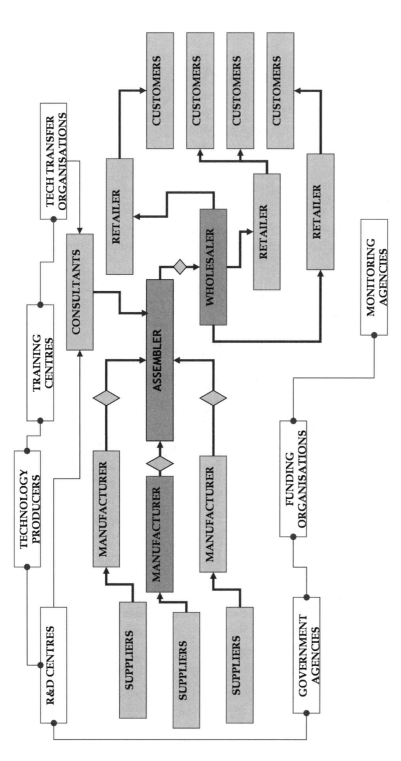

Figure 10.3 Knowledge networks within the supply chain
Partners (in rectangular) and switches (rhombi)

the partners and the flow of orders, money, information, and products. Supply chain management ensures the integration of business processes from the end-user through original suppliers who provide products, services, and information that add value for customers and other stakeholders (Patterson *et al.* 2003).

The knowledge networks related to organisational innovations follow the supply chain architecture. Products are transferred between nodes in the network; money and orders are transferred in the opposite direction; information and knowledge move in both directions. Information and knowledge networks are necessary for the functioning and optimisation of the supply chain. The partners are connected by information channels and the flow of information between two partners has to be monitored to ensure optimisation of the system.

The information and knowledge focus on products, times and costs, node performance, problems and weaknesses. This is evaluative knowledge that assists with decision-making and selecting suppliers. Major areas of knowledge are lead times, costs and quality issues, which are arranged in databases that permit comparative evaluations to be made.

Within this architecture, the institutional switches are placed between producers and suppliers and accompany ordering agreements and actual orders placed with suppliers. They serve to support evaluative judgments about suppliers (quality, cost, delivery times) and information about alternative choices among suppliers.

The combination of information and institutional switches allows the system to be managed on three basic levels: *buyer–vendor* coordination, that relates to the purchase of raw materials or parts from a single supplier or many suppliers; *production–distribution* coordination, that emphasises the integration of product manufacturing and distribution; and *inventory–distribution* coordination, that focuses on the efficiency of inventory levels to improve supply chain relationships (Chih-Ting Du *et al.* 2003).

Intelligent cities as digital spaces of collaboration

The second dimension of intelligent cities is associated to digital collaborative spaces and digital cities. Since innovation relies on knowledge and information networks, digital spaces and collaborative IT applications have become an important source of novel product, process, and organisational solutions. The main form of digital space corresponding to a territory, region or city is the digital city. Digital cities cover a very wide range of ICT networks and software applications facilitating any aspect of the social and economic life of cities: commerce, work, education R&D, transaction, security, health, leisure, and transport.

Authors of two important books on digital cities (Ishida and Isbister 2000; Tanabe *et al.* 2002) claim that the concept of digital city is a metaphor:

> As a platform for community networks, information spaces using the city metaphor are being developed worldwide.
>
> (Ishida 2000, p. 87)

> It is evident that 'digital city' is a metaphor. Metaphors (from Greek *metaphora* – transfer) serve to create new meanings by transferring the semantics of one concept into the semantics of another concept. Metaphors are habitually used to interpret an unknown 'world' (perception, experience, etc.) – the target – in terms of a familiar world – the source. Metaphorical explanation often helps us understand highly abstract and complex phenomena by relating them to phenomena we know well (or, at least, better). In so doing, a metaphor preserves (part of) the structure of the original concept, but substitutes its functional contents, anticipating the corresponding change in its properties and meaning.
>
> (Kryssanov *et al.* 2002, pp. 57–8)

This understanding is based on the assumption of a strong similarity between the physical city and its digital counterpart; a similarity that goes beyond the image of the physical space and includes structural and functional characteristics as well. The 'digital city is a metaphor called to denote a complex digital product with properties structurally similar to the ones of physical cities' (Kryssanov *et al.* 2002, p. 66).

We cannot agree with this description. It is elementary knowledge that a digital city imperfectly represents and is structurally different from the physical city of reference. All elements of the physical city do not have their equivalent digital representation. Imaginary elements may also be involved in the digital construction. Proximity in terms of distance and time is completely deformed. Even in simulations – 2D in the case of the city plan and 3D in the case of reconstruction of historical spaces and city buildings – similarity does not go beyond the form of the city. The functional aspects of the city are poorly represented by extreme simplification. Social and economic relations are not represented at all.

For us, a digital city is a collaborative digital space used to facilitate and augment the activities and functions taking place within the physical space of the city. City functions emerge from geographical concentration, infrastructure development, and cooperation within the population. Once established they provide useful services to citizens and other inhabitants: housing, work, education, health, leisure, entertainment, movement, security, information, innovation and learning, and others. Digital spaces are facilitators of these functions, and they are formed as distorted and instrumental representations of the city. We would characterise the digital representations as 'distorted' for two reasons. First, they represent a city partially, not fully and accurately, even including virtual elements that do not exist in the physical space; and second, they are mainly instrumental spaces aiming to fulfill predefined operations. Digital cities follow both the space and functions of the physical city. Their informational part links to the activities of the city; the site-seeing part represents the physical space of the city; e-market applications support commerce and transactions in the city; e-gov applications mediate in the provision of administration services, and so on and so forth. Through representations and links to physical city infrastructure and services, a digital city can inform and mediate in transactions and the provision of services in the fields of commerce, health, education, and government.

Understanding the digital city as a collaborative space of a community rather than as a representation, metaphor or simulation of the physical city implies that the architecture of digital cities is not homologous to that of physical ones; it does not derive from the physical city and its functions, but from the qualities of digital elements and the scope of their existence. The digital dimension has its own rationality; it is not just a derivative of physical space.

Ishida (2000) gives an account of the diversity of digital city architecture. He compares four different types of cities on the web, and looks at their architecture of data, form, and function:

- A commercial digital city; the digital cities created by America Online (AOL) which are structured as portals similar to 'yellow pages'. They provide local information, news, community resources, entertainment, and commerce, together with advertising local markets such as auto, real estate, employment, and health.
- A policy-driven or governmental digital city; the digital city of Amsterdam, which was created to facilitate communication between the municipal council and citizens.
- A virtual city; the virtual Helsinki, which represents the city using 3D models of buildings and public spaces, offering virtual tours and broadband communication between citizens and various service providers located in the city.
- A multi-purpose digital city; the digital city of Kyoto, in which people can get information on traffic, weather, parking, shopping, take a view of the physical environment and engage in sightseeing thought 3D models and panoramic pictures, while also having opportunities for interaction with other residents and visitors.

The architecture of these four cases varies enormously (on the same point, see Schuler 2002). In the most advanced multi-purpose and multi-functional digital city of Kyoto, the construction of the city is based on three layers. The first, which Ishida calls the 'information layer', contains data; it is a repository of raw material, html archives, real-time sensory data, media, text, and other data organised in geographical databases. The second layer is the 'interface layer', which contains maps of the city, 3D representations, city furniture, cars, buses, trains; avatars that simulate the human presence, and all the graphic design and objects that visualise the city. The third layer is the 'interaction layer' where people interact with each other, exchange information, and communicate. In the other cases (commercial city-portal, communication platform, and virtual city) architectures are simpler. The city is reduced to just a directory of urban information organised as a portal of logical and meaningful categories; as a platform for communication; as a forum giving access to municipal discussion and debate; and as an aggregate of visual data.

260 *Building blocks of intelligent cities*

Our survey at a number of digital cities on the web, found that their architecture is objective-driven, designed to fulfill the purposes of information, communication, and service delivery. However, it seems possible to devise a universal model of digital cities from which one can derive multiple combinations and alternative designs. The model can be described by a four level structure.

The first level is the *information storehouse*, a database including all digital content, in any format: texts, images, diagrams, sounds, video, and multimedia. This digital content is usually organised around some logical structure: the districts and the hierarchy of the city. The second level is the *applications* that structure the digital content and provide online services. A digital city that offers information services, e-marketplace, and e-government, includes at least three applications, which take up the tasks of combining digital content and delivering information, commercial, and governmental services. The third and upper level is the *user interface*, which includes all the webpages that users visit in order to get the services provided by the digital city. Driving a user around the different areas of the digital city, the user interface can utilise maps, 3D images, texts, and diagrams. Then, the fourth level is *administration*, a toolkit crossing the database and the applications that enables management of user rights to the applications and the digital content of the database. Administration does not mean redevelopment, but renewal of data, and control of who is entering the digital city space and for what purpose.

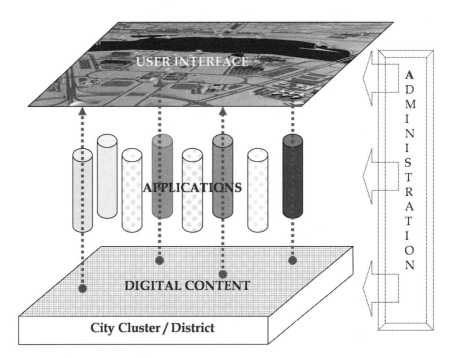

Figure 10.4 Digital cities levels

This universal architecture of digital cities is composed of three vertical levels (content, applications, and interface) and multiple horizontal functions, depending on the breadth of the digital city services (representation, information, work, leisure, commerce, transactions, etc.). The model is generic and via customisation can serve any concept of digital city specialised in site-seeing, e-government or e-work. The structure is independent of the medium on which the city runs. It may be a xDSL network, a municipal or metropolitan network made of fibre optic lines, or a local wireless network.

However, these attempts at defining the structure of digital cities with respect to a central digital space suffer from over-planning. The digital city is conceived as a fully controlled construct created by a central agency, which has absolute control over all its elements and functions. Nothing could be further from the rationale underlying the establishment of actual cities. Cities emerge rather than being planned. They arise as the result of millions of individual choices and actions, rather than as a result of a central planning and control authority.

Transferring the principles of how actual cities are organised to the field of digital and virtual cities means that it is not accurate to dub individual websites as digital cities, just like a real city is not the same as one of its buildings, no matter how large it is. On the contrary, a digital city is all websites related to the city's form, activities and functions. This applies regardless of the number of such websites or where they are located on the planet.

This concept of digital cities as the sum of digital applications generated without central planning is shown in Figure 10.5. 'Digital New York' is an example of digital space comprised of all websites turned up by an Internet search for that term on the 'Mapstan' virtual machine. This figure shows numerous websites related to NY, the similarity relationships between them since similar applications are part of the same square, and interconnections between them which are defined by the extent of links between squares.

Integration: knowledge functions of intelligent cities

Integration of territorial systems of innovation and digital collaborative spaces creates the core functions of intelligent cities. Integration occurs because knowledge networks sustaining innovation may be considerably enhanced by digital collaborative spaces. Various digital environments facilitate the establishment and operation of knowledge networks in new product development, technology transfer, and the supply chain, offering a series of knowledge storage, processing, communication, and dissemination tools.

Once knowledge networks and digital cooperative spaces are integrated, the basic knowledge functions of intelligent cities are established: strategic intelligence, technology acquisition, collaborative innovation, and global promotion. These functions may support all sectors of an intelligent city, the individual branches of industry and services, transport, the environment, and city governance. Their importance in innovation is documented by a large survey of Arthur D. Little (2005), in which more than 800 companies from across the globe

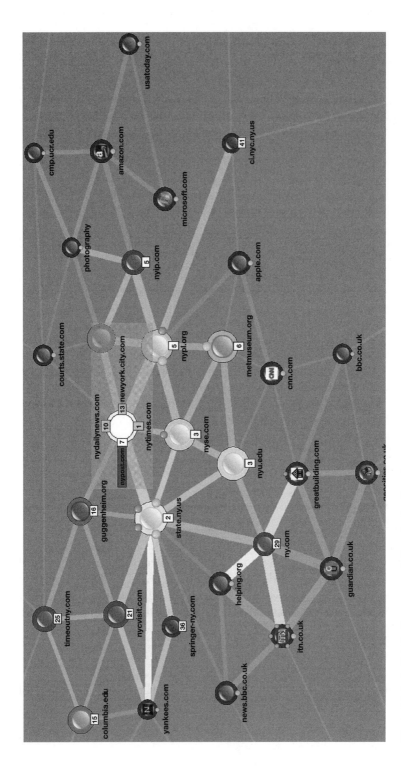

Figure 10.5 A digital city as sum of all webs for the city
Source: http://search.mapstan.net/en/splan.jsp?planId=&q=New+York&Results=-1

provided their insights on innovation excellence. According to the survey, a well-balanced innovation approach is the key to success, based on business and technology intelligence, product portfolio management, technology management, product development and launch, and post launch practices.

Let us now remind some key elements of these knowledge functions and how knowledge networks, institutional switches, and digital collaborative spaces are interconnected into the environment of intelligent cities.

Strategic intelligence

A field of innovation that has profited enormously from the information society is strategic economic intelligence. Digital cities and digital collaborative spaces may advance a particular form of strategic intelligence, 'collective strategic intelligence', in which information collection, assessment, and dissemination rely on the combined action of a group of people, a community, or a business network.

Collective strategic intelligence differs substantially from business intelligence, the best known form of economic intelligence. The latter relates to the exploitation of internal company information gathered from suppliers and customers; it uses data from enterprise resource planning (ERP) and customer relationship management (CRM), and applying data mining and data compilation techniques produces reports elucidating hidden aspects of the business environment and activity.

Collective strategic intelligence, on the contrary, is cooperative. Data comes from a group of organisations or actors, which disclose and share internal information. Information assessment is also collective and combines individual views and evaluations from the group members. Information outcomes are more robust and open to a wider information landscape.

Digital platforms facilitating collective strategic intelligence merge two types of applications: technology/market watch, and benchmarking. Technology watch is a systematic form of collection, analysis, understanding and diffusion of information concerning new product announcement, technologies, industrial statistics, performance indicators, market shares, price trends, etc. Data are stored in databases, portals, blogs and other digital repositories. Data may focus on an industry sector or a territorial entity. Benchmarking builds on this data, comparing and analysing performances and drawing lessons from the best. It has been proven to be a powerful tool for intelligence and the techniques of comparative analysis have spread out into many fields of management and policy development. Benchmarking started from companies, and has spread out to clusters, territories, and policies as well. It provides insights for any type of organisation or institution, company, R&D lab, educational institution, hospital, financing institution, etc. or collective subject, such as industry sector, cluster, region, policy and strategy as well. The methodology seeks to define the range of performance variation in any field of activity, the best performance, the distance from the best, and the practices that sustain performance. Identification of best performance and the underlying best practice are the essential pillars of any form of intelligence based on benchmarking.

264 Building blocks of intelligent cities

The added value of such digital platforms appears at various levels: use of watch and benchmarking techniques becomes easier; there is no need for special knowledge from the user perspective as the digital space guides the user during all steps of the process; the intelligence is built into the application not into the user. Data management becomes more automated. Available data are stored in a database that is constantly expanding. Selection of the comparison group happens in real time. Dissemination covers wider target groups. Most information can be offered remotely, online. The Internet feeds continuously with data, while it has become the mainstream channel for technology and market watch. By using common searching techniques the users can, easily and quickly, find critical information that might help improve their performance. The users become information providers.

Technology transfer and acquisition of technologies

Knowledge networks in technology transfer are also substantially enhanced by digital collaborative spaces. By its nature technology transfer is collaborative as it entails the transmission of know-how from one organisation to another. The knowledge that is transferred is incorporated into machines, devices, people, licenses, blue prints, prototypes, research reports, and documents. Clusters and innovative agglomerations also profit from informal forms of technology transfer, learning from others and learning from shared practices. Innovation-led development has made most companies eager consumers of technologies and research outcomes. Universities are important sources of un-exploited technology deposits and intellectual property. Digital collaborative spaces may offer substantial services in technology transfer and university, industry cooperation.

Digital platforms facilitating technology transfer are based on databases and virtual assistants. Technologies are stored in the databases and online marketplaces of technology for license are created. Organisations offering technologies input their offers and the conditions of exploitation. Users seeking solutions to their technology needs may contact the technology provider online. There is a fundamental difference, however, from patent databases, which store patent abstracts designed to protect an idea from infringement. In most cases patent databases obscure the technology, making it difficult to foresee relevant applications; the objective of a patent is to protect a technology rather than inform about technology: information disclosure is a side effect of protection. On the contrary, technology transfer platforms seek to elucidate possible uses and the application of technology in different industry sectors and activities.

Technology e-marketplaces are coupled to other online services related to technology transfer: consultative services assessing a portfolio of intellectual property; evaluation of better solutions to a given problem or need; legal assistance through the deal-making process. The objective is to digitalise the practices of technology transfer as much as possible thus enabling online interaction and technology cooperation.

Collaborative new product development

We have frequently referred to newer theories of innovation that recognise the critical role of communities and networks as a fundamental condition of innovation. Relationships within scientific communities that bridge separate fields of knowledge and technologies, complementary roles and skills along the innovation chain, and information flows among suppliers, producers, and customers, all are participatory processes feeding the knowledge networks of new product development. Relying on knowledge networks external to the company, new product development becomes truly distributed and collaborative.

Taking the innovating organisation as a point of reference, collaborative innovation networks are deployed in two directions. Backward links are intended to supply inputs from scientific research and discovery because no innovation is feasible without research inputs. Forward links are intended to respond to the needs of customers and the trends in the market, because no innovation is viable far from the market.

InnoCentive gives a measure of how large a new product development network could be; extending to all parts of the globe, and integrating hundreds of experts and organisations (see Chapter 8). Another spectacular case of collaborative product development is the new Boeing 787, which involves an extraordinary degree of collaboration between Boeing and its partners located in Japan, Russia, Italy, and the US. Boeing 787 is designed concurrently by the partners who take critical decisions on materials and electronics. All partners use the same design and collaboration software platform, called Catia, made by Dassault Systèmes S.A., and then parts are virtually 'assembled' in a computer model maintained by Boeing outside its corporate firewall, called the Global Collaboration Environment. Previously the company had to produce blueprints of the parts, which were transmitted to subcontractors to be produced; then the parts were shipped back to Boeing for assembly. The new design philosophy has radically influenced the company, which is no longer just a manufacturing company, but has been transformed to a high-end product and technologies integrator. With the adoption of the collaborative product development strategy, the company is spreading design and development costs throughout its partner network, but is also building a global product marketing and sales network (Cone 2006).

Have digital cities and online collaborative spaces really got something to offer to such networks? Do they make cooperation within the research community and the market deeper and easier? Our reply is affirmative. Digital cities can and do provide valuable tools and spaces for collaborative product development, testing, and marketing for two reasons:

- First, because knowledge that feeds new product development is distributed to a large number of partners, from R&D institutes to suppliers, subcontractors, market research and market promotion organisations, creating needs for constant communication, feed-back and knowledge integration; and

- Second, because an important part of this knowledge is tacit, communication should be continuous and direct.

Distributed new product development and promotion forces towards network structures that are barely feasible without online cooperation and interaction. The message is that apart from strategic information, digital cities can offer innovative collaborative work environments covering the entire supply chain, and enabling multiple partners and skills to intervene in real time in product design and development. The larger the network is, the more efficient novelty and problem solving capability seems to be.

Collaborative environments for product development based on digital spaces may drive to problem resolution step-by-step, through the stages of new product development for instance, or may include advanced methodologies and tools, as well as learning and experimentation through simulation. The result is a substantial improvement in innovation capability, because of collaboration and combination of know-how and skills extended over a large network of knowledge workers.

Digital marketplaces and global promotion

Marketing, promotion and e-commerce are mainstream functions of digital cities. These are the areas in which most digital cities are active. Digital promotion takes multiple forms: direct marketing, attraction of people and investments, procurement and purchasing, auctions, community and e-government services.

Innovative clusters may profit enormously from these applications. The focus is the supply chain of products and services produced by a cluster or locality. Within the supply and trade channels, digital cities have multiple added values, facilitating, enhancing, and reducing costs in all forms of transactions: logistics in the supply chain; marketing and advertising; information on policies, regulations, technical standards, and incentives; finding partners, buyers, sellers, and services (Turban *et al.* 2002). Information and knowledge networks are equally necessary for the functioning and optimisation of the supply chain. The partners are connected by information channels and the flow of information between two partners has to be monitored to ensure optimisation of the system.

The difference with individual promotion and e-commerce is that digital cities promote a cluster or locality together with its products and services. For small producers and global markets, this is an undoubted advantage; for new products in niche markets, a global market is necessary, but it cannot be reached without digital promotion.

The aforementioned four functions based on the integration between territorial systems of innovation and digital collaborative spaces are interconnected. A series of homocentric circles can be used to illustrate their connections, with strategic intelligence at the epicentre, and at successive positions applications for technology transfer, product innovation, marketing and promotion. The system creates an intelligent agglomeration/cluster, combining human, institutional, and digital resources, and optimising innovative capabilities, skills, and the global reach of the cluster.

The architecture of intelligent cities

Intelligent cities are created by fusing territorial systems of innovation and digital cities for the purpose of advancing knowledge application and innovation. The fusion is based on two objective states: (1) innovation and digital cities are both community-based processes; and (2) innovation and digital cities are both knowledge-based processes. Innovative agglomerations form the core of intelligent cities, while digital collaborative spaces and digital cities work as facilitators for innovation processes within these clusters.

The combination of innovation and broadband is quite evident in the criteria used by the Intelligent Community Forum (ICF) in selecting the Top Intelligent Communities:

> ICF has developed a list of Intelligent Community Indicators that provide the first global framework for understanding how communities and regions can gain a competitive edge in today's Broadband Economy. The Indicators demonstrate that being an Intelligent Community takes more than 'being wired'. . . . It takes a combination of . . . broadband communications . . . knowledge work . . . digital democracy . . . innovation . . . and marketing to attract new employers.
>
> (Intelligent Community Forum 2007)

Thus intelligent cities are territories with a high capacity for learning and innovation, which is based on the creativity of their population, their institutions of knowledge creation and dissemination, and their digital spatiality for world-wide communication, knowledge exchange and technology assimilation. However, the distinctive characteristic of intelligent cities is their increased performance in the field of innovation (measured by usual innovation metrics), which is sustained by a high level of ICTs, virtual networking, broadband, Internet use, and online services. In this sense, intelligent cities constitute advanced territorial systems of innovation, in which the institutional mechanisms for knowledge creation and learning are coupled with and facilitated by digital spaces and online tools for communication and knowledge management. The system is structured over three levels (L).

L1: The basic level of an intelligent city is the city's **knowledge-intensive sectors and clusters** in manufacturing and services. This level gathers the creative class of the city made up of knowledgeable and talented people, scientists, knowledge workers, artists, entrepreneurs, venture capitalists and other creative people, who determine how the workplace is organised and how the city is developing. Clusters aggregate activities, creative people, organisations, physical spaces and infrastructure. Cities aggregate multiple clusters of knowledge-intensive activities at different stages of development and maturity. Diversity is the city's distinctive character. Different 'knowledge animals' live within cities: innovative clusters of companies, science zones, knowledge zones, technology districts, technology parks, innovation poles, universities, research institutes, science labs and research teams, as well as smaller ones, like incubators, technology transfer centres, and innovation centres. The important thing is to have these elements on the spot. In this, gathering knowledge is the raw material that is then processed. People, organisations and intermediaries come together within the city, cooperating, exchanging, and applying knowledge to produce goods and services. Physical proximity and agglomeration in clusters is a condition favouring cooperation and knowledge exchange, enabling the mix of activities, network building, and the emergence of trust.

L2: The second level is comprised of **knowledge networks and innovation institutions** regulating the flow of knowledge, research, learning, and innovation. This level gathers venture capital funds, regional incentive funds, technology transfer and training centres, intellectual property agencies, spin-off support institutions, technology and marketing consultants, and all kinds of technology intermediaries. These institutions actualise and manage intangible social capital mechanisms, such as collective intelligence, cooperation in innovation, knowledge flows, and funding, which guide the complex processes of innovation within the clusters of the city. Institutions and agreements regulate how knowledge creation is funded; how consortia are established; how intellectual property is secured and distributed; what incentives are given to R&D; how small companies can tap global technology and marketing networks; and how cooperation between organisations leads to new products and services.

L3: The third level is comprised of information technology and communication infrastructures, **collaborative digital spaces, e-tools, and online services for learning and innovation**. These technologies create virtual innovation environments, based on multimedia applications, expert systems, and interactive technologies, which facilitate all processes of innovation, market and technology intelligence, technology transfer, spin-off creation, collaborative new product development, and process innovation. This is a digital working environment operating in close connection with the innovative organisations of the city and the institutions regulating knowledge and innovation.

The essence of intelligent cities is that they integrate the above three levels to work in a complementary way with each other. Knowledge-intensive activities within clusters, agreements regulating innovation, institutional switches, and digital spaces come together and enhance knowledge creation, absorption, and innovation performance. All stages of innovation profit from this integration, from intelligence and getting state-of-the-art technology, to product innovation, and product promotion. Out of the integration the main knowledge functions (F) of intelligent cities emerge. Innovation is the outcome and measure of success of cooperation, synergy and integration.

F1: Strategic intelligence. Within intelligent cities strategic intelligence is constructed by integrating a network of actors active in this field, institutional agreements regulating their cooperation, and digital spaces facilitating

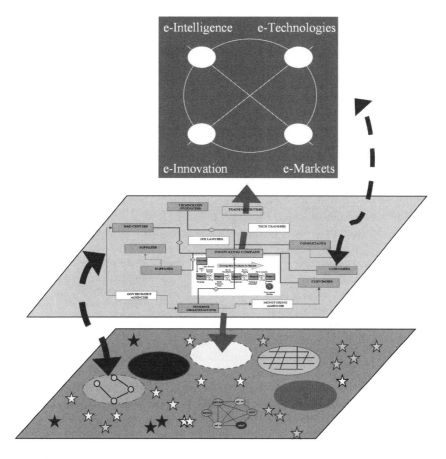

Figure 10.6 Architecture of intelligent cities

the collection, processing, and dissemination of information. The network undertakes the practices of strategic intelligence and produces an information system based on collective wisdom and interaction. The collection of information is distributed within the community, the assessment of information is also distributed and based on agreed criteria, and the dissemination of information is customised according to individual interests and needs. Development and maintenance costs are shared among the members of the community.

F2: Acquisition of technology. Absorption is related to the use of existing technologies, and technology transfer is the process of acquiring them. Getting state-of-the-art technologies is the name of the game. A series of tools that are cooperative in nature (training, demonstration, technology platforms, and technology clinics) are used to transfer these technologies and skills to the final recipients. Digital facilitators intervene in the different stages of transfer, in finding technologies from databases and libraries, online learning, and online assistance in the implementation of the technology transfer tools. An intelligent cluster actualises and brings together all elements of technology transfer: actors, transfer tools, and digital facilitators.

F3: Collaborative new product development. In collaborative innovation and new product development, actors from industry and academia set up the knowledge network. Here leadership is crucial in defining the scope and specifications of the new product. Agreements between the network members determine the role and contribution of each partner, funding, intellectual property rights, and exploitation of the product. Digital platforms assure visualisation and real-time cooperation in concept development, market testing, technical implementation, and integration of parts. This integration model is not necessarily spatially polarised. It may be used for new product development distributed on a global scale. A spatially polarised integration within intelligent clusters and cities can offer additional advantages in trust development and tacit knowledge exchange.

F4: Global promotion of localities, products and services. Promotion and sales also profit from integration of networks, agreements, and digital spaces. A group of actors corresponding to a locality or cluster promotes the cluster together with its products and services. Agreements regulate the sharing of effort and costs. Digital cities undertake marketing and e-commerce of products together with the promotion of the respective locality. Brand names are created collectively covering the group of actors and their locality or cluster. This is particularly important for smaller organisations active on global markets.

Within intelligent cities, in addition to vertical integration among levels L1, L2, and L3, a horizontal integration among functions F1, F2, F3, and F4 also takes place. Close ties link the above-mentioned functions (F1, F2, F3, and F4). Strategic intelligence is truly important for technology transfer, product innovation, and marketing. Technology transfer is a precondition for product innovation to get the best available technologies and go beyond. Marketing and promotion conclude successful product development. These functions may operate separately or in juxtaposition.

Thus the architecture of intelligent cities (gathering multiple clusters), as described, includes three levels (physical, institutional, digital) and four main functions (intelligence, absorption, innovation, and promotion). The strengths of the setting spring from the combination of individual skills, collective efforts, and intelligent machines. In fact what is integrated is human, collective, and artificial intelligence. But have no illusions. Within these collaborative innovation environments human creativity and institutional factors predominate. Digital spaces and the online expert tools act as repositories and facilitators (at least for the moment) of human and collective intelligence.

Intelligent cities: a window to global innovation networks

The way that the concept of intelligent city has been developed in this book emphasises three aspects of it:

First, its **constitution** by the superimposition of knowledge networks that emerge at different levels of the city's space: physical, institutional, and digital space. Three distinct levels comprise any intelligent city: (1) creative individuals and knowledge-intensive organisations that agglomerate in its physical space, which increasingly develop and outward-looking perspective at global opportunities and threats; (2) collective intelligence, learning, and innovation management institutions which form its institutional space; and (3) digital collaborative spaces for communication, storage and processing of information which form its digital aspect. The background underpinning the intelligent city is knowledge and in particular cooperation in the development and implementation of increasingly global knowledge. Thanks to the development of digital communication and the Internet, knowledge networks are extending across the globe, linking creative organisations from all parts of the world, while regulating principles are also becoming global. An intelligent city is a global city.

Second, its **operation** by mobilising its physical, institutional and digital networks on matters of research, information management, technology transfer, innovation, and product and service promotion. The operation of an intelligent city, or better its individual functions, arises from multiple forms of integration and collectivity in the field of knowledge. Classic city functions (housing, work, leisure, transport) have given way to another dominant group of four functions (intelligence, technology acquisition, innovation, and promotion) which manage the intangible assets of modern-day urban development. The space within which this integration takes place is global because both the knowledge and innovation networks and the digital networks operate at global level.

Third, its **results** should be clear and confirmed in terms of innovation performance. Intelligent cities have to be certified by measurable results and impact on knowledge production and application. This measurement can be made using established indicators, input and output, describing innovation drivers, knowledge generation, innovative companies, knowledge application, and intellectual property, which have been adopted by international organisations (OECD, Oslo Manual, EC, UNIDO, and others).

This emphasis on linking intelligent cities and innovation is determinative for our overall viewpoint. In different parts of this book we have argued that innovation cannot be predicted and thus modelled as a problem to be solved. We cannot systematise radical innovation because quite simply we cannot predict in what fields it will emerge. On the contrary, we can organise the environment within which innovation will occur, regardless of its special technological and production field of appearance. This is our perspective on innovation as an environmental condition. Instead of attempting to master the process of innovation itself, it is feasible to focus on improving its human, institutional and digital environment, letting creative initiatives emerge within it. We are talking about an environment made of creative people, institutions that encourage research and knowledge acquisition, cooperation, risk assumption, experimentation, the acceptance of failure, and mechanisms facilitating communication between different knowledge areas, in different branches of science and technology. In intelligent

Table 10.1 Intelligent cities building blocks and knowledge functions

	Building blocks	Knowledge functions based on the integration of Layers 1, 2, and 3
Layer 1: Physical space Physical agglomerations of innovative organisations	• Agglomerations of innovative/knowledge-intensive activities • Spontaneous clusters/technology districts • Planned clusters • Science and technology parks • Incubators • Innovation/knowledge zones • High-tech cities • Regional or sub-regional systems of innovation	
Layer 2: Institutional space Institutions, policies, and mechanisms supporting innovation	• Information collection and dissemination (Intelligence networks; Benchmarking; Futures initiatives/foresight) • Research and Development (University/R&D Labs; R&D consortia; Technology platforms) • Technology transfer (University – industry cooperation; Technology transfer centres/units; IPR management; Training) • Innovation development (New product development consortia; Global networks and alliances)	*F1: Collaborative intelligence* *F2: Technology transfer/acquisition networks* *F3: Cooperative innovation and product development*

Table 10.1 Intelligent cities building blocks and knowledge functions

	Building blocks	Knowledge functions based on the integration of Layers 1, 2, and 3
	• New company incubation (Business planning; Prototype development; Technology evaluation; Incubation consulting) • Innovation funding (VC funds; Seed funds; Business angels; Regional incentives) • Product promotion, marketing, distribution (Promotion/distribution networks) • Cluster building/vertical or horizontal cooperation (Various types of clusters and business associations)	**F4: Product promotion networks**
Layer 3: Digital space Digital spaces and web-based applications for online cooperation in innovation	• e-Intelligence (Online business/cluster intelligence; Portals/agents; Information storage; Newsletters – Visualisation – Reporting; Online benchmarking) • e-Technology (Virtual technology transfer; Virtual technology markets; Tech transfer/exploitation roadmaps; Multimedia applications) • e-Innovation (Online collaboration in innovation; Learning roadmaps; New product development tools) • e-Incubation (Virtual incubation; Business planning tools; Marketing planning tools; Market research tools; Cost-benefit analysis) • e-Marketplaces (Digital marketplaces; Virtual city tour; e-market places; e-government shops)	

cities this environment is developed and operates in the physical, institutional and digital space. It shapes an integrated, global system of innovation. Key aspects of this have been identified in the information management, existing knowledge and technology acquisition, new product development, product and service promotion, linking skills and opportunities anywhere in the world they are available.

In July 2006 I presented the architecture of intelligent cities at a conference on 'Intelligent Environments 2006' held in Athens, as superimposed knowledge networks and institutional switches over three levels (the physical, institution and digital) sustaining four knowledge functions (strategic intelligence, technology transfer, collaborative NPD and promotion on the global market). The spatiality of this city is global both in terms of its establishment and sphere of operation. However, its special feature lies in its ability to link and integrate different forms of intelligence: human intelligence, the collective intelligence of institutions regulating knowledge flows, and the artificial intelligence of digital spaces (Komninos 2006). I hope this perspective will provide a springboard for many branches of academic research that contribute to the development of intelligent cities (urban development, planning, geography, innovation management, telecommunications and IT) in constructing more complex and functional applications. The architecture of intelligent cities with their three spatial levels and four knowledge functions may be particularly useful in solving complex problems that require mobilisation of a city's population, such as improved competitiveness, the development of new technology districts, regenerating rundown areas, and running new technologies infrastructure and networks.

Today intelligent cities, communities and clusters are an attractive prospect, a strategy, and a vision for the future, rather than an actuality that has been realised. There are agglomerations which open routes towards physical–digital intelligent systems, but the road to be travelled before the emergence of truly intelligent cities and regions is long. An important step towards achieving this plan is to understand and describe the linkages between the physical, institutional, and digital aspects of intelligent environments and how those interconnections actualise creativity practices that transform knowledge into new products and novelties.

What intelligent cities can achieve are more sophisticated systems of innovation enabling, through the digital interaction, an extension of collaboration networks at global scale and the participation of users. These are two novel elements (global innovation networks/user participation to innovation) that broadband communication and digital collaborative spaces can offer to local/regional systems of innovation. The key effect of Intelligent Communities, Smart Cities, Living Labs, and other forms of intelligent environments is to enlarge the knowledge processes of the respective system of innovation with the participation of overseas suppliers, innovators, and end-users. They do it through intense networking, both at local and global scales.

Appendix
Five platforms for intelligent cities

Creating intelligent cities is not exclusively a digital technology problem. Digital applications are important and vital, but need to be linked into the knowledge-intensive clusters that operate in the city, the skills and specialisation of the population, the institutions for generating and managing knowledge, and the intermediary organisations promoting innovation. Making an intelligent city is tied into improving skills among a city's population, planning innovative clusters, innovation centres, technology transfer centres, technology parks, and setting intangible networks that foster knowledge and innovation. It is against this background that digital tools and virtual innovation spaces can operate.

As a means of facilitating the design of intelligent cities, we developed at URENIO a series of digital platforms that provide guidance on the creation of the core knowledge functions of these cities. The platforms support five key innovation processes: strategic economic intelligence, technology dissemination, collaborative innovation and new product development, digital cities and e-marketplaces, and new companies incubation.

For each process a separate platform has been created containing information management tools, AI applications for alert, search, information classification, processing, and dissemination, which are addressed to global collaborations and end-users. The philosophy of the platforms is to actualize both global innovation networks and web 2.0 applications for the participation of users.

The platforms help in creating the internal knowledge processes of an intelligent city for resolving multiple challenges: developmental, environmental or social. They stand on the assumption that any problem can be solved following the sequence: 'acquisition of accurate information' – 'adoption of state-of-the-art technology' – 'development of a new solution' – 'promotion of the solution'. The platforms offer assistance in implementing this approach.

Five platforms for intelligent cities

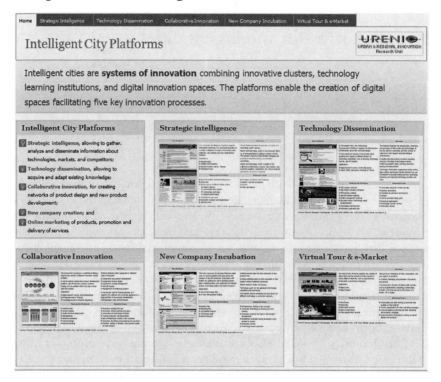

Figure A.1 Five platforms for intelligent cities
Source: www.urenio.org/platforms/index.html

The virtual innovation environments created by these platforms enable each district of a city/region (productive cluster, technology district, central-city area of services, technology park, incubator, university campus, or other area of information-intensive activities) to deploy its knowledge functions using broadband networks, digital cooperation spaces, and online services. The objective is not the broadband per se, but the opening of the district's knowledge processes to global collaboration and the participation of the users. Broadband services and virtual environments are just the medium for making the systems of innovation more open and user responsive.

Strategic economic intelligence platform

Economic intelligence is the process of gathering and utilising information for business and development purposes. The term describes the process of turning data into information and then into knowledge. The intelligence is claimed to be more useful to the user as it passes through each step. The strategic economic intelligence platform supports both the needs of individual companies, as well as needs of industry sectors and clusters. The platform is structured according to usual business intelligence principles, including data collection, data analysis, and data dissemination modules.

The knowledge model behind the platform has two core components. The first focuses on collecting, evaluating, and storing data concerning markets and technologies. Watch services are designed and implemented on a per sector basis. The second component focuses on data analysis and reporting. The main analysis tool is benchmarking. Benchmarking of different fields such as industry sectors, commodities and markets, enterprise performance, communities, cities, and regional performance, can give meaning to data and provide insights on better performance.

The platform supports intelligence at the business, sector, and regional level. Thus potential users are companies, organisations managing and promoting industries, clusters, business associations, chambers of industry and commerce, and regional authorities and development associations as well. More generally, users are those organisations that consider market and technology watch and the systematic follow-up of emerging trends as an important component for their activity and competitiveness.

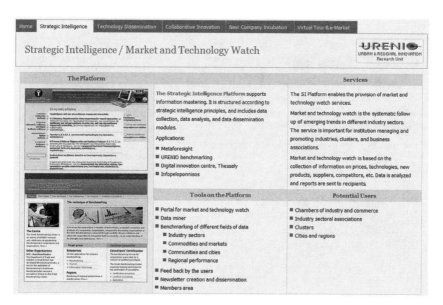

Figure A.2 Strategic economic intelligence platform
Source: www.urenio.org/platforms/si.html

Technology dissemination/acquisition platform

The platform for R&D dissemination and exploitation serves as a hub for connecting technology supply and demand. It offers the necessary tools that facilitate the dissemination, marketing, and promotion of R&D and technology services. It includes three modules, each of which contains a number of e-tools and applications:

- Online R&D database: Technology providers from universities and R&D centres submit information about research, technologies, products and services; technology users can access this information over the web.
- Online innovation learning: Technology training and learning are based on roadmaps that guide R&D exploitation. They offer self-training and help users to accomplish tasks such as technology transfer, spin-off company creation, IP management, and conclusion of contracts.
- Online collaboration: Collaboration between academia and businesses is achieved through the use of online communication tools: a technology-matching tool, a discussion forum, and a cooperation space. These tools create a digital space where entrepreneurs, intermediaries, and public organisations can post their technology needs/problems, which are then automatically communicated to the closest technology provider, and then work together with researchers and experts on the solution of the problem.

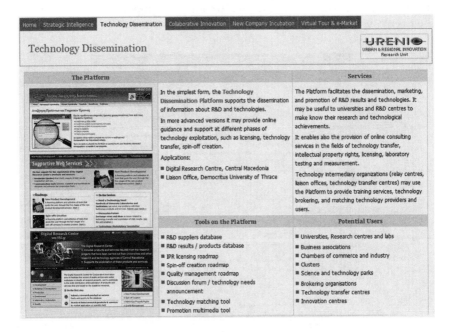

Figure A.3 Technology dissemination platform
Source: www.urenio.org/platforms/td.html

Collaborative innovation platform

New product development (NPD) is at the core of business innovation and the Collaborative Innovation Platform is designed both to support learning and to facilitate NPD.

The platform offers a set of tools that can assist companies and organisations to successfully develop new products or upgrade existing ones through a series of logical steps, starting from idea generation and ending with the launch of the product. It contains a series of *Levels* and control points, called *Assessments*. Each *Level* contains information and well-defined e-tools concerned with a particular phase of product development. Each *Assessment* is a decision point where senior management can keep on with or stop the process. Within this structure, the platform links to a series of tools useful during the product development process, such as conjoint analysis, quality functional deployment, brainstorming, reverse engineering, industrial design, rapid prototyping, and others.

The platform can be used not only as a learning tool, but also as a complete guide to NPD. Potential users are diverse: companies developing new products; incubators assisting start-ups and monitoring new product development of tenants; innovation centres assisting small companies; training companies and organisations as a value added service to their clients; new entrepreneurs starting a new company and more directly, scientists wishing to develop a new product based on their research.

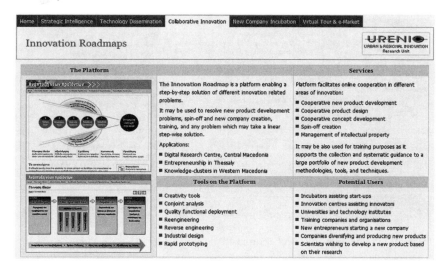

Figure A.4 Collaborative innovation platform
Source: www.urenio.org/platforms/ci.html

New company incubation platform

This platform helps users to resolve typical problems that arise during the creation of spin-off companies. It is complementary to the Technology Dissemination Platform since spin-off creation is a principal technology dissemination process. The platform was initially developed to assist technology transfer operations in technology parks, but it evolved as a toolbox for business planning.

The platform contains a full toolbox guiding the fundamental operations that starting companies deal with: business planning, marketing, assessment of alternative technologies to use, market research. Dedicated tools facilitate each operation with templates to fill out, Excel sheets for calculation, control checks, interpretation of data and facts. At the end, e-tools automatically produce complete business and marketing plans, cost-benefit analyses, and market research guides.

Potential users of the platform are entrepreneurs starting a new company, academics wishing to start a new business and commercialise their research, but mostly incubators and incubatees.

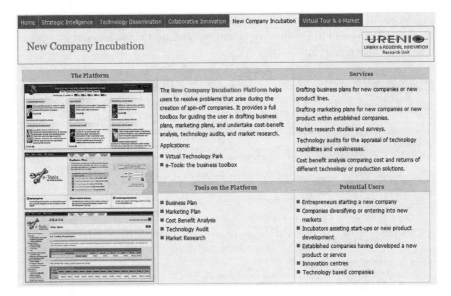

Figure A.5 New company incubation platform
Source: www.urenio.org/platforms/nci.html

Virtual tour and e-marketplace platform

The platform facilitates the design of digital cities and the promotion of cities together with local products and services. It enables the provision of services in various fields of urban life, such as e-government, e-promotion, and e-business. The platform contains three modules:

- The virtual city, which provides virtual tours of the city, presenting monuments, art crafts, points of interest with the use of digital maps and panoramic photographs.
- The e-marketplace where companies and business clusters can provide information, present offers, carry out e-commerce, form business relationships, etc.
- The e-government shop, which provides the seamless aggregation of service provision in the public administration realm.

Figure A.6 Virtual tour and e-marketplace platform
Source: www.urenio.org/platforms/dcs.html

The virtual city component acts as a main medium for the city's promotion in tourism and cultural terms. By showing places of interest (monuments, sights, public buildings, infrastructures, education and research facilities, recreation areas, etc.) on the map and providing relevant cultural information it helps the city's residents or visitors organise their visit or spare time according to their special interests.

The e-marketplace enables companies and citizens to create e-shops within the e-marketplace and offer products and services over the Internet. Individual e-shops are placed and connected to the virtual city representation.

The e-government shop is primarily operated by the city's municipal authority. It allows users to report a wide range of problems and queries, to apply online for many municipal services and certificates, and to pay for council services online payment. The e-government module can also provide information regarding the activities of local government or municipal authorities.

Potential users of the platform are communities and cities wishing to promote their places on the Internet; tourism organisations, for their marketing campaigns; local and regional authorities aiming to deliver their services online; local associations of producers wishing to market digitally their products taking advantage of local brand names.

Notes

1. The Oslo Manual is an initiative of the OECD, the European Commission, and Eurostat focusing on the measurement of innovation and providing a framework within which existing surveys in OECD countries can evolve towards comparability. The first version of the Oslo Manual was issued in 1992. The surveys undertaken using it, mainly the Community Innovation Surveys (CIS), showed that it is possible to develop and collect data on the complex and differentiated process of innovation. The Manual has set the standard for the European Innovation Scoreboard, which is published each year since 2001.
2. This well-know tale is about a turkey on a farm. The first day she spent on the farm she was fed at 9 a.m. However, as a good deductionist, she did not draw hasty conclusions. She waited to assemble a larger number of observations that confirmed that they fed her at 9 a.m. She gathered observations under various circumstances: Wednesdays and Thursdays, warm and cold days, rain and sun. Each day she added one more observation on her list. When finally her deductionist conscience was satisfied and she considered that she was in a position to advance a deductive conclusion, she stated: 'they feed me each day at 9 a.m.'. Unfortunately, this conclusion was to be proved false in the most indisputable way, when on the Christmas Eve, instead of feeding her at 9 a.m., as she expected, they slaughtered her (Chalmers, 1994).
3. We have here an explanation very close to the way Becattini (1989) explained how an industrial district works as a creative milieu.
4. As the recent restructuring of Airbus – via the Power 8 plan – showed, production decentralisation is universal. It does not only relate to so-called traditional sectors (textiles, clothing, metals, etc.) but also new technology, high added value sectors such as aerospace. Thanks to Power 8, which is being implemented at a time of high profits and a market upturn, the percentage of added value by Airbus of the A 350 aircraft is around 50 per cent (at the same level as Boeing's Dreamliner) by transferring production departments to Korea, India and Russia (Ruffin 2007).

References

Abramovsky, L., Jaumandreu, J., Kremp, E., and Peters, B. (2004) 'National differences in innovation behaviour – facts and explanations results using basic statistics from CIS 3 for France, Germany, Spain and United Kingdom'. Available online: www.eco.uc3m.es/IEEF/ieef-cis3.pdf (accessed 1 September 2007).

The Age (2007) 'Ten capitals of the world' (18 June 2007). Available online: www.theage.com.au/news/technology/tech-capitals-of-the-world/2007/06/16/1181414598292.html?page=fullpage#contentSwap2 (accessed 4 September 2007).

Albert, S. (2006) *Smarten Up: A Guide to Creating a Smart Community*, Victoria: Trafford.

Allen, R. (1998) 'The web: interactive and multimedia education', *Computer Networks and ISDN Systems*, 30: pp. 1717–27.

Antonelli, C. (2000) 'Collective knowledge communication and innovation: the evidence of technological districts', *Regional Studies*, 34, 6: pp. 535–47.

Anttiroiko, A.V. (2004) 'Science cities: their characteristics and future challenges', *International Journal of Technology Management*, 28 (3–4–5–6): pp. 395–418.

Archibugi, D. and Coco, A. (2004) 'International partnerships for knowledge in business and academia. A comparison between Europe and the USA', *Technovation*, 24 (7): pp. 517–28.

Arora, A., Fosfuri, A., and Gambardella, A. (2002) *Markets for Technology: The Economics of Innovation and Corporate Strategy*, Cambridge, MA: MIT Press.

ARTE (2003) 'The top 10 critical challenges for business intelligence success: White Paper', *ComputerWorld*. Available online: www.atre.com/bi_whitepaper (accessed 5 July 2004).

Arthur D. Little (2005) 'Innovation excellence 2005: How companies use innovation to improve profitability and growth'. Available online: www.adlittle.nl/insights/studies/innovation_excellence_study_2005.php (accessed 11 October 2007).

Atlee, T. (2005) 'Definitions of collective intelligence', *Blog of Collective Intelligence*. Available online: www.community-intelligence.com/blogs/public/archives/000288.html#more (accessed 10 March 2005).

Azzone, G. and Maccarrone, P. (1997) 'The emerging role of lean infrastructures in technology transfer: the case of the Innovation Plaza project', *Technovation*, 17 (7): pp. 391–402.

Back, T. (2002) 'Adaptive business intelligence based on evolution strategies: some application examples of self-adaptive software', *Information Sciences – An International Journal*, 148: pp. 113–21.

Bagnasco, A. (1977) *Tre Italia. La problematica territoriale dello sviluppo economico italiano*, Bologna: Il Mulino.

Batty, M. (2005), *Cities and Complexity: Understanding Cities with Cellular Automata: Agent-based Models and Fractals*, Cambridge, MA: MIT Press.

Becattini, G. (1979) 'Dal settore industriale al distretto industriale. Alcune considerazioni sull' unita di indagine dell'economia industriale', *Rivista di Economia e Politica industriale*, 5: pp. 7–21.

Becattini, G. (1989) 'Le district industriel: milieu creatif', *Espaces et Societes, Revue Scientifique Internationale*, 66–7: pp. 147–63.

Becattini, G. (1991) 'The industrial district as a creative milieu', in M. Dunford and G. Benko (eds), *Industrial Change and Regional Development*, London: Belhaven Press.

Becker, W. and Dietz, J. (2004) 'R&D cooperation and innovation activities of firms – evidence for the German manufacturing industry', *Research Policy*, 33 (2): pp. 209–25.

Beckman (2004) 'Intelligence', Beckman Institute for Advanced Science and Technology. Available online: www.beckman.uiuc.edu/research/bioint.html (accessed 10 March 2004).

Bell, R. (2006) 'Intelligent communities: leveraging IT for economic and social growth', Intelligent City Forum. Available online: www.intelligentcommunities.org (accessed 6 August 2007).

Bertacchini, Y. and Dou, H. (2001) 'The territorial competitive intelligence: a network concept'. Available online: http://archivesic.ccsd.cnrs.fr/documents/archives0/00/00/04/36/sic_00000436_02/sic_00000436.html (accessed 10 December 2007).

Besselaar, P. and Koizumi, S. (eds) (2005) *Digital Cities III: Information Technologies for Social Capital – Cross-cultural Perspectives*, Berlin: Springer-Verlag.

Besselaar, P. and Beckers, D. (2005) 'The life and death of the Great Amsterdam Digital City', in P. Besselaar and S. Koizumi (eds), *Digital Cities III: Information Technologies for Social Capital – Cross-cultural Perspectives*, Berlin: Springer-Verlag, pp. 66–96.

Blomqvist, K., Hara, V., Koivuniemi, J., and Äijö, T. (2004) 'Towards networked R&D management: the R&D approach of Sonera corporation as an example', *R&D Management*, 34 (5): pp. 591–603.

Bobrowski, P. (2000) 'A framework for integrating external information into new product development: lessons from medical technology industry', *Journal of Technology Transfer*, 25: pp. 181–92.

Bonaccorsi, A. and Lipparini, A. (1994) 'Strategic partnerships in new product development: an Italian case study', *Journal of Product Innovation Management*, 11: pp. 134–45.

Booz Allen Hamilton (2006) 'The four sources of intelligent innovation: winning the race for profitable growth'. Available online: www.leading-innovations.com/lic05-home/lic03 (accessed 21 February 2007).

Bounfour, A. and Edvinsson, L. (eds) (2005) *Intellectual Capital for Communities: Nations, Regions, and Cities*, Oxford: Elsevier, Butterworth Heinemann.

Bowen-James, A. (1997) 'Paradoxes and parables of intelligent environments', in P. Droege (ed.), *Intelligent Environments – Spatial Aspect of the Information Revolution*, Oxford: Elsevier, Butterworth Heinemann.

Braczyk, H., Cooke, P., and Heidenreich, R. (eds) (1997) *Regional Innovation Systems*, London: UCL Press.

Breschi, S. (2000) 'The geography of innovation: a cross-sector analysis', *Regional Studies*, 34 (3): pp. 213–29.

Brusco, S. (1982) 'The Emilian model: productive decentralisation and social integration', *Cambridge Journal of Economics*, 6: pp. 167–84.

BusinessWeek (2003) 'The future of technology', *BusinessWeek*, European Edition, Special Issue (August 18–25), pp. 33–100.

BusinessWeek (2005) 'Get creative', *BusinessEweek*, Special Report, 1 August 2005. Available online: www.businessweek.com/magazine/content/05_31/b3945401.htm (accessed 28 July 2006).

BusinessWeek (2007) 'Inside innovation – special report', *BusinessWeek*, 10 August 2007. Available online: www.businessweek.com/innovate/di_special/20070830insideinnov.htm (accessed 19 September 2007).

Caillard, D. (2007) 'L'Intelligence collective'. Available online: http://barthes.ens.fr/scpo/Presentations00-01/Caillard_IntelligenceCollective/intcol.htm (accessed 8 July 2007).

California Institute for Smart Communities (2001) *Smart Communities Guidebook*. Available online: www.smartcommunities.org/index2.html (accessed 7 September 2007).

Caloghirou, Y., Ioannides, S., Tsakanikas, A., and Vonortas, N., (2004) 'Subsidized research joint ventures in Europe', in Y. Caloghirou, N. Vonortas, and S. Ioannides (eds), *European Collaboration in Research and Development*, Cheltenham: Edward Elgar.

Calvin, W.H. (1998) *How Brains Think. Evolving Intelligence, Then and Now*, London: Phoenix.

Camagni, R. (1991) (ed.) *Innovation Networks: Spatial Perspectives*, London, Belhaven.

Cantwell, J.A. (1989) *Technological Innovation and Multinational Corporations*, Oxford: Basil Blackwell.

Carayannis, E., Assimakopoulos, D., and Kondo, M. (eds) (2007) *Innovation Networks and Knowledge Clusters: Findings and Insights from the US, EU and Japan*, Basingstoke: Palgrave Macmillan.

Cargo, R. (2000) 'Made for each other: non-profit management education, online technology, and libraries', *The Journal of Academic Librarianship*, 26 (1): pp. 15–20.

Carrillo, F.J. (ed.) (2005) *Knowledge. Cities: Approaches, Experiences and Perspectives*, Burlington, MA: Elsevier, Butterworth-Heinemann.

Castells, E. and Cardoso G. (eds) (2006) *The Network Society: From Knowledge to Policy*, Washington, DC: Center for Transatlantic Relations.

Castells, M. (1996) *The Rise of the Network Society*, Oxford: Blackwell.

CETISME (2002) 'Economic intelligence: a guide for beginners and practitioners', Brussels: European Communities. Available online: www.madrimasd.org/Quees madrimasd/Socios_Europeos/descripcionproyectos/Documentos/CETISME-ETI-guide-english.pdf (accessed 13 September 2007).

Chalmers, A. F. (1994) *What Is This That We Call Science?* Heracleion: Crete University Press.

Chesbrough, H. (2003) *Open Innovation: The New Imperative for Creating and Profiting from Technology*, Boston, MA: Harvard Business School Press.

Chiesa, V., Manzini, R., and Pizzurno, E. (2004) 'The externalisation of R&D activities and the growing market of product development services', *R&D Management*, 34 (1): pp. 65–75.

Chih-Ting Du, T., Lee H. M., and Chen, A. (2003) 'Constructing federated databases in coordinated supply chains', *Decision Support Systems*, 36: pp. 49–64.

Choo, C.W. (1997) 'IT 2000: Singapore's vision of an intelligent island', in P. Droege (ed.), *Intelligent Environments: Spatial Aspects of the Information Revolution*, New York: Elsevier, pp. 49–65.

Choy, K.L., Lee, W.B., and Lo, Victor (2003) 'An intelligent supplier management system for selecting and benchmarking suppliers', *International Journal of Technology Management*, 26 (7): pp. 717–42.

Churchill, E., Girgensohn, A., Nelson, L., and Lee, A. (2004) 'Blending digital and physical spaces for ubiquitous community participation', *Communications of the ACM*, 47 (2) (February): pp. 39–45.

Clausen, T.H. (2004) 'Technological regimes and Norwegian manufacturing', Centre for Technology, Innovation and Culture (TIK), University of Oslo. Available online: www.druid.dk/uploads/tx_picturedb/ds2004-1346.pdf (accessed 31 August 2007).

Cohen, J., Hart, D., and Simmie J. (1997) (eds) *Recherche et Developpement Regional*, Paris: Publications de la Sorbonne.

Cohen, M. (1997) 'Towards interactive environments: the intelligent room', Proceedings of the 1997 Conference on Human Computer Interaction, Bristol, United Kingdom. Available online: http://people.csail.mit.edu/mhcoen/Papers/aaai98.pdf (accessed 7 September 2007).

The Collective Intelligence Lab (2007) 'Collective intelligence'. Available online: http://137.122.100.152 (accessed 8 July 2007).

Cone, E. (2006) 'Boeing: new jet, new way of doing business', *CIO Inside*. Available online: www.cioinsight.com/article2/0,1540,1938904,00.asp (accessed 11 October 2007).

Cooke, P. (1996) 'The new wave of regional innovation networks: analysis, characteristics and strategy', *Small Business Economics*, 8: pp. 1–13.

Cooke, P. (2003) 'Economic globalisation and its future challenges for regional development', *International Journal of Technology Management*, 26 (2/3/4): pp. 401–20.

Cooke, P. and Morgan, K. (1998) *The Associational Economy: Firms, Regions and Innovation*, Oxford: Oxford University Press.

Cooke, P. and Wills, D. (1999) 'Small firms, social capital and the enhancement of business performance through innovation programmes', *Small Business Economics*, 13: pp. 219–34.

Cooke, P., Uranga-Gomez, M., and Extebarria, G. (1997) 'Regional innovation systems: institutional and organisational dimensions', *Research Policy*, 26: pp. 475–91.

Cooke, P., Uranga-Gomez, M., and Extebarria, G. (1998) 'Regional systems of innovation: an evolutionary perspective', *Environment and Planning A*, 30 (9): pp. 1563–84.

Cooper R.G. (1999) 'New product development', in M.J. Baker (ed.), *International Encyclopedia of Business & Management: Encyclopedia of Marketing*, 1st edn, London: International Thomson Business Press, pp. 342–55.

Cooper, R.G. (1994) 'Third generation new product process', *Journal of Product Innovation Management*, 11: pp. 3–14.

Cooper, R.G. (1996) 'Overhauling the new product process', *Industrial Marketing Management*, 25: pp. 465–82.

Cooper, R.G. and Edgett, S.J. (2007) 'Stage gate process', Product Development Institute Inc. Available online: http://prod-dev.com/stage-gate.shtml (accessed 1 August 2007).

Cordis (2003) 'Regions of knowledge'. Available online: www.cordis.lu/era/knowreg.htm (accessed 13 September 2003).

Cordis (2007) 'Science and technology indicators for the European research area (STI-ERA)'. Available online: http://cordis.europa.eu/indicators (accessed 12 May 2005).

Corsi, P., Christofol, H., Richir, S., and Samier, H. (2007), *Innovation Engineering: The Power of Intangible Networks*, London: ISTE.

Crang, P. and Martin, R. (1991) 'Mrs Thatcher's vision of the new Britain and the other sides of the Cambridge phenomenon', *Environment and Planning D: Society and Space*, 9 (1): pp. 91–116.

Crow, K. (2005) 'Collaboration'. Available online: www.npd-solutions.com/collaboration.html (accessed 17 July 2007).

Culley, S. (1999) 'Suppliers in new product development: their information and integration', *Journal of Engineering Design*, 10 (1): 59–75.

D'Agostino, G. (2000) 'Regional patterns of innovation: the analysis of CIS results and lessons from other innovation surveys', Final Report. Available online: ftp://ftp.cordis.lu/pub/eims/docs/eims_summary_98_192.pdf (accessed 1 September 2007).

Datta, A. and Mohtadi, H. (2007) 'Endogenous imitation and technology absorption in a model of north-south trade'. Available online: www.uwm.edu/~mohtadi/Datta-Mohtadi-NS-Trade-90-10-05.pdf (accessed 17 July 2007).

Davenport T.H. and Prusak, L. (1998) *Working Knowledge. How Organizations Manage What They Know*, Boston, MA: Harvard Business School Press.

Dawes, P. (2003) 'A model of the effect of technical consultants on organizational learning in high-technology purchase situations', *Journal of High Technology Management Research*, 14: pp. 1–20.

Digital Birmingham (2007) 'Creating the digital city of the future today'. Available online: www.digitalbirmingham.co.uk (accessed 6 August 2007).

Dosi, G., Freeman, C., Nelson, R., Silverberg, G., and Soete, L. (eds) (1988) *Technical Change and Economic Theory*, London and New York: Frances Pinter.

Dou, H. (2000) 'De l'information a l'intelligence. Les procédés de création d'intelligence compétitive pour action dans les PME et PMI', Conférence A3F, Sophia Antipolis, October 2000.

Droege, P. (ed.) (1997) *Intelligent Environments – Spatial Aspect of the Information Revolution*, Oxford: Elsevier, Butterworth-Heinemann.

East Midlands Observatory (2007). Available online: www.eastmidlandsobservatory.org.uk/index.asp (accessed 8 July 2007).

East of England Observatory (2007). Available online: www.eastofenglandobservatory.org.uk (accessed 8 July 2007).

Economist Intelligence Unit (2007) 'Sharing the idea: the emergence of global innovation networks'. Available online: www.idaireland.com/home/news.aspx?id=9&content_id=656 (accessed 28 August 2007).

Edquist, C. (ed.) (1997) *Systems of Innovation: Technologies, Institutions and Organizations*, London and Washington: Frances Pinter.

Edquist, C. and Lundvall, B.-A. (1993) 'Comparing the Danish and Swedish systems of innovation', in R. Nelson and N. Rosenberg (eds), *National Systems of Innovations: A Comparative Analysis*, Oxford: Oxford University Press, pp. 265–98.

Edquist, C., Rees, G., Lorenzen, M., and Vincent-Lancrin, S. (2001) *Cities and Regions in the Learning Economy*, Paris: OECD Publications.

Edvinsson, L. (2005) 'Regional intellectual capital in waiting: a strategic intellectual capital quest', in A. Bounfour and L. Edvinsson (eds), *Intellectual Capital for Communities, Nations, and Cities*, Oxford: Elsevier.

Eger, J. (1997) 'Cyberspace and cyberplace: building the smart communities of tomorrow', *San Diego Union-Tribune*, 26 October 1997. Available online: www.smartcommunities.org/library_cyberspace.htm (accessed 5 July 2007).

EIS (2001–2006) European Innovation Scoreboards 2001–2006, 'European trend chart on innovation', Cordis. Available online: http://trendchart.cordis.lu/tc_innovation_scoreboard.cfm (accessed 1 September 2007).

Enright, M. (2000) 'The globalization of competition and the localization of competitive advantage: policies toward regional clustering', in N. Hood and S. Young (eds),

Globalization of Multinational Enterprise Activity and Economic Development, London: Macmillan.

Ettlie, J. (2006) *Managing Innovation: New Technology, New Products, and New Services in a Global Economy*, Amsterdam and Boston, MA: Elsevier, Butterworth-Heinemann.

European Commission (1994) *Competitiveness and Cohesion: Trends in the Regions, Fifth Periodic Report on the Social and Economic Situation and Development of the Regions in the Community*, Luxembourg: Office for Official Publications of the European Communities.

European Commission (1999) *Sixth Periodic Report on the Social and Economic Conditions of the European Regions*, Luxembourg: Official Publications of the European Communities.

European Commission (2000) *The Regions in the New Economy: Guidelines for Innovative Actions under the ERDF in the Period 2000–6*, Communication from the Commission to the member states. Available online: http://ec.europa.eu/regional_policy/sources/docoffic/official/guidelines/pdf/inovac_en.pdf (accessed 7 September 2007).

European Commission (2001a) *Second Report on Economic and Social Cohesion*, Luxembourg: Office of Official Publications of the European Communities.

European Commission (2001b) *Innovation Statistics in Europe: Data 1996–97*, Luxembourg: Official Publications of the European Communities.

European Commission (2001c) *European Competitiveness Report 2001*, Luxembourg, Office of Official Publications of the European Communities. Available online: http://europa.eu.int/comm/enterprise/enterprise_policy/competitiveness/doc/competitiveness_report_2001 (accessed 1 September 2007).

European Commission (2002a) 'The European research area: providing new momentum, strengthening, reorienting, opening up new perspectives', Communication from the Commission, Brussels, Com (2002) 556.

European Commission (2002b) *Innovation Policy in Europe 2001*, Luxembourg: Official Publications of the European Communities.

European Commission (2003) 'Innovation policy: updating the Union's approach in the context of the Lisbon strategy', Communication from the Commission to the Council, the European Parliament, the European Economic and Social Committee and the Committee of the Regions, Brussels, Com (2003) 112.

European Commission (2004a) *European Competitiveness Report 2004*, Luxembourg: Office of Official Publications of the European Communities.

European Commission (2004b) *Third Report on Economic and Social Cohesion. The Socio-economic Situation of the Union and the Impact of European and National Policies*, Luxembourg: Office of Official Publications of the European Communities. Available online: http://europa.eu/scadplus/leg/en/lvb/g24006.htm (accessed 21 June 2005).

European Commission (2005) 'A new start for the Lisbon strategy'. Available online: http://europa.eu/scadplus/leg/en/cha/c11325.htm (accessed 29 November 2006).

European Commission (2007) *Growing Regions, Growing Europe. Fourth Report on Economic and Social Cohesion*, Luxembourg: Office of Official Publications of the European Communities. Available online: http://ec.europa.eu/regional_policy/sources/docoffic/official/reports/cohesion4/index_en.htm (accessed 26 August 2007).

Eurostat (2007) 'Survey on innovation in EU enterprises'. Available online: http://europa.eu.int/estatref/info/sdds/en/inn/inn_base.htm (accessed 1 September 2007).

Evangelista R., Sandven, T., Sirilli, G., and Smith, K. (1998) 'Measuring innovation in European industry'm *International Journal of the Economics of Business*, 5 (3): 311–33.

Feldmann, M. and Muller, S. (2003) 'An incentive scheme for true information providing in supply chains', *Omega*, 31: pp. 63–73.

Flash Eurobarometer (2002) 'Innobarometer 2002', No. 129. Available online: http://ec.europa.eu/public_opinion/flash/fl129_en.pdf (accessed 10 September 2007).

Fleischer, M. and Liker, J. (1997) *Concurrent Engineering Effectiveness – Integrating Product Development Across Organizations*, Cincinnati, OH: Hanser Gardner Publications.

Florida, R. (2002) *The Rise of the Creative Class and How It's Transforming Work: Leisure, Community and Everyday Life*, New York: Basic Books.

Florida, R. (2005) *The Flight of the Creative Class: The New Global Competition for Talent*, New York: Harper Business.

Fogel, D. B. (1995) *Evolutionary Computation: Towards a Philosophy of Machine Intelligence*, Piscataway, NJ: IEEE Press.

Freeman, C. (1987) *Technology Policy and Economic Performance: Lessons from Japan*, London: Pinter Publishers.

Freeman, C. (1991) 'Networks of innovators: a synthesis of research issues', *Research Policy*, 20: pp. 499–514.

Freeman, C. (1995) 'The "national system of innovation" in historical perspective', *Cambridge Journal of Economics*, 19: pp. 5–24.

Friedman, T.L. (2006) *The World is Flat: The Globalized World in the Twenty First Century*, London: Penguin Books.

Garavelli, A.C., Gorgoglione, M., and Scozzi, B. (2002) 'Managing technology transfer by knowledge technologies', *Technovation*, 22: pp. 269–79.

Gavigan, J., Keenan, M., Miles, I., Fahri, F., Lecoq, D., Capriati, M., and Di Bartolomeo, T. (eds) (2001) *A Practical Guide to Regional Foresight*, FOREN Network, European Commission, DG Research.

Gentler, M. (1996) 'Barriers to technology transfer: culture and the limits to regional systems of innovation', Paper presented at the Conference RESTPOR 96, Brussels, 19–21 September.

Gibson, W. (1984) *Neuromancer*, New York: ACE Books.

Gloor, P. (2006) *Swarm Creativity: Competitive Advantage Through Collaborative Innovation Networks*, Oxford: Oxford University Press.

Goldense, B. and Schwartz, A. (2002) 'When companies outsource R&D, the main focus is NPD', *Product Development and Management Association Visions Magazine*. Available online: www.pdma.org/visions/july04/outsourcing.html (accessed 2 October 2007).

Gottinger, H. (2006) *Innovation, Technology and Hypercompetition*, London: Routledge.

Graf, H. (2007), *Networks in the Innovation Process: Local and Regional Interactions*, Cheltenham: Edward Elgar.

Graham, S. (ed.) (2003) *The Cybercities Reader*, London: Routledge.

Griliches, Z. (1979) 'Issues in assessing the contribution of R&D to productivity growth', *Bell Journal of Economics*, 10: pp. 92–116.

Griliches, Z. (ed.) (1984) *R&D Patents and Productivity*, Chicago, IL: University of Chicago Press.

GSRT (2005) 'Regional innovation poles in Greece'. Available online: www.gsrt.gr (accessed 10 June 2005).

Hagedoorn, J., Link, A.N., and Vonortas, N. (2000) 'Research partnerships', *Research Policy*, 29: pp. 567–86.

Hall, I. and Hardy, S. (2003) 'RSA annual conference. Building entrepreneurial capacity in the regions', *Regions*, No. 243.

Hansson, F., Husted K., and Vestergaard, J. (2005) 'Second generation science parks: from structural holes jockeys to social capital catalysts of the knowledge society', *Technovation*, 26 (9): pp. 1039–49.

Hargadon, A. (2003) *How Breakthroughs Happen: The Surprising Truth About How Companies Innovate*, Boston, MA: Harvard Business School Press.

Harmon, B., Ardishvili, A., Cardoso, R., Elder, T., Leuthold, J., Parshall, J., Raghian, M., and Smith, D. (1997) 'Mapping the university technology transfer process', *Journal of Business Venturing*, 12: pp. 423–34.

Heidenreich, M. (2004) 'The dilemmas of regional innovation systems', in P. Cooke, M. Heidenreich, and H.J. Braczyk (eds), *Regional Innovation Systems*, London: Routledge.

Henri, F. and Pudelko, B. (2003) 'Understanding and analysing activity and learning in virtual communities', *Journal of Computer Assisted Learning*, 19: pp. 474–87.

Herbaux, Ph. and Chotin, R. (2002) 'L'intelligence économique, outil du pacte territorial', Paper presented in the ASRDLF Conference, University of Quebec, 21 April.

Heylighen, F. and Joslyn, C. (2001) 'Cybernetics and second order cybernetics', in R.A. Meyers (ed.), *Encyclopedia of Physical Science & Technology* (3rd edn), New York: Academic Press.

Holbrook, J.A.D. (1997) 'The use of national systems of innovation models to develop indicators of innovation and technological capacity', CPROST Report # 97–06, Centre for Policy Research on Science and Technology, Vancouver: Simon Fraser University at Harbour Centre.

Hollanders, H. (2006) *2006 European Innovation Scoreboard*, Brussels: European Commission, DG Enterprise.

Holmes, J.S. and Glass, J.T. (2004) 'Internal R&D – vital but only one piece of the innovation puzzle', *Research Technology Management*, 47 (5): pp. 7–22.

Howells, J. (1999) 'Research and technology outsourcing', *Technology Analysis and Strategic Management*, 11 (1): pp. 17–29.

Iansiti, M. (1993) 'Real world R&D. Jumping the product development gap', in *Managing High-Tech Industries, Harvard Business Review*, Vol. 71 (3): pp. 138–47.

Intelligent Community Forum (2001) 'The top seven intelligent communities of 2001'. Available online: www.intelligentcommunity.org (accessed 5 July 2001).

Intelligent Community Forum (2003) 'What is an intelligent community?' Available online: www.intelligentcommunity.org/displaycommon.cfm?an=1&subarticlenbr=18 (accessed 10 March 2004).

Intelligent Community Forum (2006) 'Intelligent community indicators'. Available online: www.intelligentcommunity.org/displaycommon.cfm?an=1&subarticlenbr=4 (accessed 5 July 2006).

Intelligent Community Forum (2007) 'Intelligent community awards'. Available online: www.intelligentcommunity.org/displaycommon.cfm?an=1&subarticlenbr=54 (accessed 6 July 2007).

Index of Silicon Valley (2007). Available online: www.jointventure.org/inthenews/pressreleases/2007index.html (accessed 28 August 2007).

Innocentive (2007). Available online: www.innocentive.com/using/index.html (accessed 1 August 2007).

INTA – Annual Congress (2007) 'INTA30 world urban development congress: creativity, competitiveness and community'. Available online: www.inta-aivn.org/index.php?option=com_content&task=section&id=5&Itemid=83 (accessed 7 September 2007).

Intelligent Environment Lab (2007) 'Intelligent environment project'. Available online: www.cs.memphis.edu/~tmccauly/html/intelligent_environment_projec.html (accessed 20 August 2007).

Intelligent Environments 07 (2007) 'Conference outline'. Available online: www.uni-ulm.de/ie07 (accessed 21 June 2007).

Ishida, T. and Isbister, K. (eds) (2000) *Digital Cities: Technologies, Experiences, and Future Perspectives*, Berlin: Springer-Verlag.

Ishida, T. (2000) 'Understanding digital cities', in T. Ishida and K. Isbister (eds), *Digital Cities: Experiences, Technologies and Future Perspectives, Lecture Notes in Computer Science*, Vol. 1765, Heidelberg: Springer-Verlag.

Ishida, T. (2000) 'Understanding digital cities', in T. Ishida and K. Isbister (eds), *Digital Cities: Experiences, Technologies and Future Perspectives, Lecture Notes in Computer Science*, Berlin: Springer-Verlag.

Ishida, T. and Isbister, K. (eds) (2000) *Digital Cities: Experiences, Technologies and Future Perspectives, Lecture Notes in Computer Science*, Vol. 1765, Berlin and Heidelberg: Springer-Verlag.

Jaffe, A.B. (1986) 'Technological opportunity and spillovers of R&D: evidence from firms', *American Economic Revue*, 76: 984–1001.

Jaffe, A.B. (1989) 'Real effects of academic research', *American Economic Revue*, 79: 957–70.

Jaffe, A.B., Trajtenberg, M., and Henderson, R. (1993) 'Geographic location of knowledge spillovers as evidenced by patent citations', *Quarterly Journal of Economics*, 63: 577–98.

Jaruzelski, B., Dehoff, K., and Bordia, R. (2006) 'Smart spenders: global innovation 1000', *Strategy and Business Magazine*, Winter.

John Adams Innovation Institute (2007) 'The index of the Massachusetts innovation economy'. Available online: www.masstech.org/institute/the_index/archive.htm (accessed 8 July 2007).

Johnson, A. (2006) 'CI 2.0: competitive innovation intelligence', KMWorld 2006, San Jose, California, 13 October. Available online: www.slideshare.net/arikjohnson/ci-20-competitive-innovation-intelligence (accessed 14 July 2007).

Juniper, J. (2002) 'Universities and collaboration within complex, uncertain knowledge-based economics', *Critical Perspectives on Accounting*, 13: 747–78.

Kafkalas, G. and Komninos, N. (1999) 'The innovative region strategy: lessons from the C. Macedonia regional technology plan', in K. Morgan and C. Neuwelaers (eds), *Regional Innovation Strategies and Peripheral Regions*, London: J. Kingsley.

Kaufmann, A. and Todtling, F. (2000) 'Systems of innovation in traditional industrial regions: the case of Styria in a comparative perspective', *Regional Studies*, 34 (1): 29–40.

Keeble, D. and Wilkinson, F. (1998) 'Collective learning and knowledge development in the evolution of regional clusters of high tech SMEs in Europe', *Regional Studies*, 334: 295–303.

Keeble, D., Lawson C., Moore, B., and Wilkinson, F. (1998) 'Collective learning processes, networking and institutional thickness in the Cambridge region', *Regional Studies*, 33 (4): 319–32.

Keeble, D., Lawson, C., Lawton-Smith, H., Moore, B., and Wilkinson, F. (1998) 'Internationalisation processes, networking and local embeddedness in technology-intensive small firms', *Small Business Economics*, 11: 327–42.

Keeble, D., Lawson, C., Moore, B., and Wilkinson, F. (1999) 'Collective learning processes, networking and institutional thickness in the Cambridge region', *Regional Studies*, 33 (4): 319–32.

Kelessidis, V. (2000) 'Benchmarking'. Available online: www.urenio.org/benchmark/reports.asp (accessed 6 October 2004).
Kelly, J. (2005) 'Wireless in the city'. Available online: http://searchnetworking.techtarget.com/originalContent/0,289142,sid7_gci1091714,00.html (accessed 5 October 2007).
Kennedy, J. and Eberhart, R. (2001) *Swarm Intelligence*, San Francisco, CA: Academic Press.
Kessler, E. (2003) 'Leveraging e-R&D processes: a knowledge based view', *Technovation*, 23: pp. 905–15.
Kimzey, C. and Kurokawa, S. (2002) 'Technology outsourcing in the US and Japan', *Research – Technology Management*, 45 (4): pp. 36–42.
Kitsios, F. (2005) 'Innovation management and new product development', Ph.D. dissertation, Technical University of Crete.
Knowledge Board (2004) 'Agent and multi-agent systems for knowledge management'. Available online: www.knowledgeboard.com/cgi-bin/item.cgi?id=1262 (accessed 5 September 2007).
Komninos, N. (2004) 'Regions of excellence in the EU: a new model of regional hierarchy and development', in G. Kafkalas (ed.), *Spatial Development Issues*, Athens: Kritiki Press.
Komninos, N. (2002) *Intelligent Cities: Innovation, knowledge systems and digital spaces*, London and New York: Spon Press.
Komninos, N. (2004) 'Regional intelligence: distributed localized information systems for innovation and development', *International Journal of Technology Management*, 28 (3–4–5–6): pp. 483–506.
Komninos, N. (2006) 'The architecture of intelligent cities', *Intelligent Environments 06*, Institution of Engineering and Technology, 53–61. Available online: www.urenio.org/el/wp-content/uploads/2007/07/the-architecture-of-intel-cities-ie06.pdf (accessed 7 September 2007).
Komninos, N., Sefertzi, E., and Tsarchopoulos P. (2006) 'Virtual innovation environment for the exploitation of R&D', *Intelligent Environments 06 Proceedings*, Institution of Engineering and Technology, pp. 125–36.
Krugman, P. (1991) *Geography and Trade*, Palatino Cambridge, MA: MIT Press.
Kryssanov, V.V., Okabe, M., Kakusho, K., and Minoh, M. (2002) 'Communication of social agents and the digital city: a semiotic perspective', in M. Tanabe, P. Van den Besselaar, and T. Ishida (eds), *Digital Cities*, LNCS No. 2362, Berlin, Heidelberg: Springer-Verlag.
Kuhlmann, S., Boekholt, P., Georghiou, L., Gyu, K., Heraud, J., Laredo, P., Lemola, T., Loveridge, D., Luukkonen, T., Polt, W., Rip, A. Sanz-Menendez, L., and Smits, R. (1999) 'Improving distributed intelligence in complex innovation systems', Final Report, TSER project, Fraunhofer Institute, Systems and Innovation Research, Karlsruhe. Available online: www.isi.fraunhofer.de/p/Final.pdf (accessed 6 July 2007).
Kyrgiafini, L. and Sefertzi, E. (2003) 'Changing regional systems of innovation in Greece: the impact of regional innovation strategy initiatives in peripheral areas of Europe', *European Planning Studies*, 11 (8): pp. 885–910.
Lall, S. (2007) 'Technology absorption: an overview'. Available online: http://siteresources.worldbank.org/INTECAREGTOPKNOECO/Resources/SanjayaLall.ppt (accessed 20 July 2007).
Lambert, D. M. and Cooper, M.C. (2000) 'Issues in supply chain management', *Industrial Marketing Management*, 29: pp. 65–83.

Landabaso, M. (1999) 'EU policy on innovation and regional development' in F. Boekema, K. Morgan, S. Bakkers, and R. Rutten (eds) *Knowledge, Innovation and Economic Growth: The Theory and Practice of Learning Regions*, London: Edward Elgar Publishing.

Landabaso, M. (2007) 'The regional economic development relevance of social capital', in M. Landabaso, A. Kuklinski, and C. Roman (eds), *Social Capital, Innovation and Regional Development: The Ostuni Consensus*, Nowy Sacz: WSB-National Louis University.

Lawson, C. and Lorenz, E. (1999) 'Collective learning, tacit knowledge and regional innovative capacity', *Regional Studies*, 33 (4): pp. 305–17.

Lee J. and Win, H.N. (2004) 'Technology transfer between university research centres and industry in Singapore', *Technovation*, 24: pp. 433–42.

Lévy, P. (1996) 'The second flood – report on cyberculture', conference on 'A New Space for Culture and Society (New Ideas in Science and Art)', Prague, 19–23 November 1996. Available online: www.culturelink.org/review/21/cl21levy.html (accessed 5 October 2007).

Lévy, P. (1994) *L'Intelligence Collective: Pour une anthropologie du cyberspace*, Paris: La Découverte.

Lévy, P. (1997) *Collective Intelligence: Mankind's Emerging World in Cyberspace*, Cambridge, MA: Perseus Books.

Lévy, P. (2007) 'Le futur Web exprimera l'intelligence collective de l'humanité'. Available online: www.pointblog.com/past/000232.htm (accessed 8 July 2007).

Lichtenthaler, U. and Lichtenthaler, E. (2004) 'Alliance functions: implications of the international multi-R&D-alliance perspective', *Technovation*, 24 (7): pp. 541–52.

Lim, A. (2001) 'Intelligent island discourse: Singapore's discursive negotiation with technology', *Bulletin of Science, Technology and Society*, 21 (3): pp. 175–92.

Limthanmaphon, B., Zhang, Z., and Zhang, Y. (2004) 'Intelligent web-based e-commerce system' in T. Shih and P. Wang (eds), *Intelligent Virtual World: Technologies and Applications in Distributed Virtual Environment*, London: World Scientific, pp. 295–323.

Linturi, R. and Simula, T. (2005) 'Virtual Helsinki: enabling the citizen – linking the physical and virtual', in P. Besselaar and S. Koizumi (eds), *Digital Cities III: Information Technologies for Social Capital – Cross-cultural Perspectives*, Berlin: Springer-Verlag, pp. 113–40.

Living Labs Europe (2007) 'Living Labs Europe: a new driver for European innovation'. Online www.livinglabs-europe.com/livinglabs.asp (accessed 5 July 2007).

Love, J. and Roper, S. (2002) 'Internal versus external R&D: a study of R&D choice with sample selection', *International Journal of the Economics of Business*, 9 (2): pp. 239–55.

Love, J. and Roper, S. (2004) 'The organisation of innovation: collaboration, cooperation and multifunctional groups in UK and German manufacturing', *Cambridge Journal of Economics*, 28: pp. 379–95.

Lovering, J. (1999) 'Theory led by policy: the inadequacies of the "new regionalism" (illustrated from the case of Wales)', *International Journal of Urban and Regional Research*, 23: pp. 379–95.

Lundvall, B. and Johnson, B. (1994) 'The learning economy', *Journal of Industry Studies*, 1: pp. 23–41.

Lundvall, B. (ed.)(1992) *National Systems of Innovation: Towards a Theory of Innovation and Interactive Learning*, London: Pinter Publishers.

McLeod, G. (2001) 'New regionalism reconsidered: globalisation and remaking of political economic space', *International Journal of Urban and Regional Research*, 25: 804–29.

Macpherson, A. (1997) 'The contribution of external service inputs to the product development efforts of small manufacturing firms', *R&D Management*, 27 (2): pp. 127–42.

MADRI+D (2007). Available online: www.madrimasd.org (accessed 8 July 2007).

MalarNetCity (2007) 'What is MalarNetCity'. Available online: www.malarnetcity.se/texter/read.php?id=99317 (accessed 6 August 2007).

Malerba, F. (2002) 'New challenges for sectoral systems of innovation in Europe', DRUID Summer Conference on Industrial Dynamics of the New and Old Economy, Copenhagen, 6–8 June 2002. Available online: www.druid.dk/uploads/tx_picturedb/ds2002-630.pdf (accessed 1 September 2007).

Malone, R. (2005) 'Disruptive technologies'. Available online: www.managingautomation.com/maonline/magazine/read.jspx?id=2326569&printable (accessed 21 June 2007).

Markusen, A. (1996) 'Sticky places in slippery spaces: a typology of industrial districts', *Economic Geography*, 72: pp. 293–313.

Marshall, G.L. (2000) 'Virtual communities and their network support: a cybernetic analysis', *Cybernetics and Systems: An International Journal*, 31: pp. 397–415.

Maskell, P. and Malmberg, A. (1995) 'Localised learning and industrial competitiveness', Paper presented at the Regional Studies Association European Conference on Regional Futures, Gothenburg, 6–9 May 1995.

Masser, I. (1990) 'Technology and regional development policy: a review of Japan's Technopolis programme' *Regional Studies*, 24 (1): pp. 41–53.

Mazower, M. (2004) *Salonica City of Ghosts: Christian, Muslims, and Jews 1430–1950*, London: Happer Perennial.

Metcalfe, S. (1995) 'The economic foundations of technology policy: equilibrium and evolutionary perspectives', in P. Stoneman (ed.), *Handbook of the Economics of Innovation and Technological Change*, Oxford: Blackwell Publishers.

MetroFi (2007) 'Fast, free, wireless internet for everyone'. Available online: www.metrofi.com (accessed 6 August 2007).

Miles, I., Keenan, M., and Clar, G. (eds) (2002) *Practical Guide to Regional Foresight in the United Kingdom*, Seville: European Commission – Joint Research Centre – Institute for Prospective Technological Studies

Miller, W. and Morris, L. (1999) *Fourth Generation R&D: Managing Knowledge, Technology, and Innovation*, New York: Wiley.

Millson, M.R. and Raj, S.P. (1996) 'Strategic partnering for developing new products', *Research and Technology Management*, 69 (3): pp. 41–50.

Minsky, M. (1988) *The Society of Mind*, New York: Simon & Schuster.

Miotti, L. and Sachwald, F. (2003) 'Cooperative R&D: why and with whom? – an integrated framework of analysis', *Research Policy*, 32: pp. 1481–99.

Mitchell, J.W. (2000) 'Designing the digital city' in T. Ishida and K. Isbister (eds), *Digital Cities: Technologies, Experiences, and Future Perspectives*, Berlin: Springer-Verlag, pp. 1–6.

Moreno, R., Paci, R., and Usai, S. (2005) 'Geographical and sectoral clusters of innovation in Europe', *Annals of Regional Science*, 39: pp. 715–39.

Morgan, K. (1997) 'The learning region: institutions, innovation and regional renewal', *Regional Studies*, 31 (5): pp. 491–503.

Morgan, K. (2001) 'The exaggerated death of geography: localised learning, innovation and uneven development', Paper presented to The Future of Innovation Studies Conference, The Eindhoven Centre for Innovation Studies, Eindhoven University of Technology, 20–3 September 2001.

Morgan, K. (2004) 'The exaggerated death of geography: learning, proximity and territorial innovation systems', *Journal of Economic Geography*, 4: pp. 3–21.

Morse, S. (2004) *Smart Communities: How Citizens and Local Leaders Can Use Strategic Thinking to Build a Brighter Future*, New York: Jossey-Bass.

MuniWireless (2007) 'St Cloud Florida citywide Wi-Fi update: lunch plus 30 days'. Available online: www.muniwireless.com/article/articleview/5118 (accessed 6 August 2007).

Narula, R. and Hagedoorn, J. (1999) 'Innovating through strategic alliances: moving towards international partnerships and contractual agreements', *Technovation*, 19: pp. 283–94.

Nelson, R. and Rosenberg, N. (eds) (1993) *National Innovation Systems: A Comparative Analysis*, Oxford: Oxford University Press.

Nelson, R. and Winters, S. (1982) *An Evolutionary Theory of Economic Change*, Cambridge, MA: Harvard University Press.

Newman, P. and Thornley, A. (2004), *Planning World Cities: Globalization, Urban Governance and Policy Dilemmas*, Basingstoke: Palgrave Macmillan.

Nonaka, I. and Takeuchi, H. (1995) *The Knowledge-Creating Company*, Oxford: Oxford University Press.

North, K. and Kares, S. (2005) 'Ragusa or how to measure ignorance: the ignorance meter', in A. Bounfour and L. Edvinsson (eds), *Intellectual Capital for Communities, Nations, Regions, and Cities*, Oxford: Elsevier, Butterworth-Heinemann, pp. 253–64.

Novak, M. (1997) 'Cognitive cities: intelligence, environment and space', in P. Droege (ed.), *Intelligent Environments – Spatial Aspect of the Information Revolution*, Oxford: Elsevier, Butterworth-Heinemann.

Novak, S. and Eppinger, S. (2001) 'Sourcing by design: product complexity and the supply chain', *Management Science*, 47 (1): pp. 189–204.

Onecommunity (2007) 'Connecting – enabling – transforming'. Available online: www.onecleveland.org (accessed 5 July 2007).

Oslo Manual (1995) *The Measurement of Scientific and Technological Activities*, OECD and Eurostat. Available online: www.oecd.org/dataoecd/35/61/2367580.pdf (accessed 1 September 2007).

Ozer, M. (2003) 'Process implications of the use of internet in new product development: a conceptual analysis', *Industrial Marketing Management*, 32: 517–30.

Paci, R. and Usai, S. (2000) 'Technological enclaves and industrial districts. An analysis of regional distribution of innovative activities in Europe', *Regional Studies*, 34 (2): pp. 97–114.

Passiante, G. and Secundo, G., (2002) 'From geographical innovation clusters towards virtual innovation clusters: the innovation virtual system', Paper presented in the 42nd ERSA Congress, From Industry to Advanced Services – Perspectives of European Metropolitan Region, Dortmund, 27–31 August 2002.

Patel, P. and Pavitt, K. (1994), 'The nature and economic importance of national innovation systems', *STI Review*, No. 14, Paris: OECD.

Patterson, K.A., Grimm, C.M., and Corsi, T.M. (2003) 'Adopting new technologies for supply chain management', *Transportation Research*, 39: pp. 95–121.

Pawar, B.S. and Sharda, R. (1997) 'Obtaining business intelligence on the Internet', *Long Range Planning*, 30 (1): pp. 110–21.

Piergiovanni, R. and Santarelli, E. (2001) 'Patents and the geographic localization of R&D spillovers in French manufacturing', *Regional Studies*, 35 (8): pp. 697–702.

Piore, M. and Sabel, C. (1984) *The Second Industrial Divide: Possibilities for Prosperity*, New York: Basic Books.

Pittaway, L., Robertson, M., Munir, K., Denyer, D., and Neely, A. (2004) 'Networking and innovation: a systematic review of the evidence' *International Journal of Management Reviews*, 5–6 (3–4): pp. 137–68.

Polard D. (2005) 'Innovation as collaboration'. Available online: http://blogs.salon.com/0002007/2005/01/26.html#a1031 (accessed 2 October 2007).

Poles de Competitivite (2007) 'Les Poles de Competivite'. Available online: www.competitivite.gouv.fr (accessed 6 July 2007).

Porter, M. (1990) *The Competitive Advantage of Nations*, New York: Free Press.

Poti, B. and Basile, R. (2000) 'Differences in innovation performance between advanced and backward regions in Italy: the role of firms' strategies, organizational factors and institutions', Rome: ISRDS/CNR and IASE.

Pugh O'Mara, M. (2004) *Cities of Knowledge: Cold War Science and the Search for the Next Silicon Valley*, New Jersey: Princeton University Press.

Putnam, R. (1995) 'Turning in, turning out: the strange disappearance of social capital in America', *Political Science and Politics*, 28 (4): pp. 664–79.

Quazzotti, S., Dubois, C., and Dou, H. (1999) 'Veille technologique, guide des bonnes pratiques en PME et PMI', Report to the European Commission, DG XIII, The Innovation Programme, Luxembourg.

Quinn, J.B. (2000) 'Outsourcing innovation: the new engine for growth', *Sloan Management Review*, 41 (4): pp. 13–28.

R&D Magazine, (2002) 'Outsourcing in times of change', *R&D Magazine*, 12: pp. 29–30.

Radjou, N. (2004) 'Innovation networks: a new market structure will revitalize invention-to-innovation cycles'. Available online: www-03.ibm.com/technology/businessvalue/files/innovation-networks.pdf (accessed 12 November 2004).

Radovanovic, D. (2003) 'Intelligence and Lund: what lessons Lund can learn in order to become an intelligent city', MA thesis, University of Lund, School of Economics and Management. Available online: www.entovation.com/whatsnew/Intelligence_Lund.pdf (accessed 10 October 2007).

Raison, D. (1998) 'Intelligence territoriale: le cas du Poitou-Charentes', in *NET 98: Le salon de l'Internet et de l'Intranet*. Available online: http://cat.inist.fr/?aModele=afficheN&cpsidt=1814870 (accessed 6 August 2007).

Regional Intelligence Unit (2007). Available online: www.nwriu.co.uk (accessed 8 July 2007).

Reid, A. (2003) 'Ongoing evaluation report of the VERITE network: interim report', Brussels: Reid Consulting.

Reid, A. (2004) 'Evaluation report of the VERITE network: final report', Brussels: Reid Consulting.

Reiterer, H., Mubler, G., Mann, T., and Handschuh, S. (2000) 'INSYDER – an information assistant for business intelligence', Proceedings of the 23rd annual international ACM SIGIR conference on research and development in information retrieval, Athens, pp. 112–19. Available online: http://portal.acm.org/citation.cfm?id=345559 (accessed 6 August 2007).

Republique et Canton de Geneve (2007) 'E-voting'. Available online: www.geneve.ch/evoting/english/welcome.asp (accessed 6 August 2007).

Rogers E. M., Takegami, S., and Yin J. (2001) 'Lessons learnt about technology transfer', *Technovation*, 21: pp. 253–61.

Rouibah, K. and Ould-ali, S., (2002) 'PUZZLE: aa concept and prototype for linking business intelligence to business strategy', *Strategic Information Systems*, 11: pp. 133–52.

Ruffin, F. (2007) 'Airbus on the land, stocks on the air', *Le Monde Diplomatique*, 17 June.

Santos, J., Doz, Y., and Williamson, P. (2004) 'Is your innovation process global', *MIT Sloan Management Review*, 45 (4): pp. 31–7.

Saperstein, J. and Rouach, D. (2002) *Creating Regional Wealth in the Innovation Economy*, New Jersey: Financial Times Prentice Hall.

Sassen, S. (2003) 'Agglomeration in the digital era', in S. Graham (ed.), *Cybercities Reader*, London: Routledge, pp. 191–9.

Sassen, S. (2006) *Cities in a World Economy*, Thousand Oaks, CA: Pine Forge Press.

Saxenian, A. (1990) 'Regional networks and the resurgence of Silicon Valley', *California Management Review*, Fall: pp. 89–111.

Schuler, D. (2002) 'Digital cities and digital citizens', in M. Tanabe, P. Van den Besselaar, and T. Ishida (eds), *Digital Cities, Lecture Notes on Computer Science*, No. 2362, Berlin and Heidelberg: Springer-Verlag, pp. 71–85.

Schumpeter, J.A. (1934) *The Theory of Economic Development*, Cambridge MA: Harvard University Press.

Schumpeter, J.A. (1943) *Capitalism, Socialism and Democracy*, London: Allen & Unwin.

Schwen, T. and Hara, N. (2003) 'Community of practice: a metaphor for online design', *The Information Society*, 19: pp. 257–70.

The Scottish Parliament (2007) 'Public petitions committee: e-petitions'. Available online: http://epetitions.scottish.parliament.uk/default.asp (accessed 6 August 2007).

Scott, A. (1988a) 'Flexible production systems and regional development: the rise of new industrial spaces in North America and Western Europe', *International Journal of Urban and Regional Research*, 12 (2): pp. 171–86.

Scott, A. (1988b) *New Industrial Spaces. Flexible Production, Organisation and Regional Development in North America and Western Europe*, London: Pion.

Selby, D. (2002) 'Jotting from the business intelligence jungle', Proceedings of the 2002 conference on APL: array processing languages, lore, problems, and applications, Madrid, pp. 190–7.

Sethi, R., Pant, S., and Sethi, A. (2003) 'Web-based product development systems integration and new product outcomes: a conceptual framework', *Journal of Product Innovation Management*, 20: pp. 37–56.

Shah, K. and Sheth A. (2000) 'Logical information modelling of web-accessible heterogeneous digital assets'. Available online: http://lsdis.cs.uga.edu/lib/download/SS98.htm (accessed 9 September 2007).

Shih, T. and Wang, P. (eds) (2005), *Intelligent Virtual World: Technologies and Applications in Distributed Virtual Environment*, London: World Scientific.

Shiode, N. (1997) 'An outlook for urban planning in cyberspace: toward the construction of cyber cities with the application of unique characteristics of cyberspace', International Symposium on City Planning, Nagoya, Japan, September 1997.

Simmie, J. (1997) *Innovation, Networks and Learning Regions?* London: Jessica Kingsley.

Simmie, J. (2001) (ed.) *Innovative Cities*, London: Spon Press.

Simmie, J. and Sennett, J. (1999) 'Innovative clusters: theoretical explanations and why size matters', *National Institute Economic Review*, 170: pp. 87–98.

Simovits, M. and Forsberg, T. (1997) 'Business intelligence and information warfare on the Internet', Paper presented at INFOSEC'COM International Congress on Information Systems and Telecommunications Security, 12–13 June 1997, CNIT, Paris: La Defence.

Smart Communities (2007) 'The World Foundation for Smart Communities'. Available online: www.smartcommunities.org/about.htm (accessed 6 July 2007).

Smeds, R. and Alvesalo, J. (2003) 'Global business process development in a virtual community of practice', *Production and Planning Control*, 14 (4): pp. 361–71.

Springer, R. (2005) 'InnoCentive hosts the world's largest virtual laboratory', *India West*, XXX (26).

Sternberg, R. (1998) 'Innovative linkages and proximity: empirical results from recent surveys of small and medium sized firms in Germany regions', *Regional Studies*, 33 (6): pp. 529–40.

STEP Economics (2000) 'The analysis of CIS II data: towards an identification of regional innovation systems', Final Report, Turin. Available online: ftp://ftp.cordis.europa.eu/pub/eims/docs/eims_summary_98_192.pdf (accessed 30 August 2007).

Steventon, A. and Wright, S. (eds) (2006) *Intelligent Spaces: The Application of Pervasive ICT*, Berlin, Heidelberg: Springer-Verlag.

Stiroh, K.J. (2002) 'Information, technology and the US productivity revival: what do the industry data say?', *American Economic Review*, 92 (5): pp. 1559–76.

Storper, M. (1997) *The Regional World*, London and New York: Guilford Press.

Storper, M. and Scott, A. (1988) 'The geographical foundations and social reproduction of flexible production complexes', in J. Wolch and M. Dear (eds), *Territory and Social Reproduction*, London: Allen & Unwin.

Strombach, S. (2001) 'Innovation clusters and innovation processes in the Stuttgart region', in J. Simmie (ed.), *Innovation Cities*, London: Spon Press, pp. 53–95.

Szuba, T. (2001) *Computational Collective Intelligence*, New York: Wiley.

Tanabe, M., Van den Besselaar, P., and Ishida, T. (eds) (2002) *Digital Cities: Lecture Notes in Computer Science*, No. 2362, Berlin and Heidelberg: Springer-Verlag.

Taylor, P., Derudder, B., Saey, P., and Witlox, F. (2006) *Cities in Globalisation: Practices, Policies and Theories*, London: Routledge.

Tether, B. (2002) 'Who cooperates in innovation, and why: an empirical analysis', *Research Policy*, 31: pp. 947–67.

Tidd, J. (1995) 'Development of novel products through intra-organisational and inter-organisational networks – the case of home automation', *Journal of Product Innovation Management*, 12: pp. 307–22.

Tidd, J., Bessant, J., and Pavitt, K. (2001) *Managing Innovation: Integrating Technological, Market and Organizational Change*, New York: Wiley.

Tsoukalas, I. and Anthopoulos, L. (2004) 'Moving towards the e-city'. Available online: www.centerdigitalgov.com/international/story.php?docid=92186 (accessed 5 July 2007).

Turban, G., Mclean, E., and Wetherbe, J. (2002) *Information Management. Transforming Business in the Digital Economy*, New York: John Wiley & Sons, Wiley International Edition.

URENIO (2007a) 'Meta-foresight guide', Brussels. Available online: www.urenio.org/metaforesight/guide.html (accessed 9 September 2007).

URENIO (2007b) 'Netforce: ten technology transfer manuals'. Available online: www.intelspace.eu/kmt (accessed 10 September 2007).

Van den Besselaar, P. and Koizumi, S. (2005), *Digital Cities III. Information Technologies for Social Capital: Cross-cultural Perspectives*, Berlin: Springer-Verlag.

Veille (2002) 'Dossier DECiLOR', *Le Magazine Professionel de l'Intelligence Economique et du Management de la Connaissance*, No. 50, pp. 10–20, November.

Verona, G., Prandelli, E., and Sawhney, M. (2006) 'Innovation and virtual environments: towards virtual knowledge brokering', *Organization Studies*, 27: pp. 765–88.

Von Hippel, E. (1988) *The Sources of Innovation*, Oxford: Oxford University Press.

Von Hippel, E. (2002) 'Innovation by user-communities: learning from open-source software', in E. Roberts (ed.), *Innovation: Driving Product, Process and Market Change*, San Francisco, CA: Jossey-Bass.

The Wall Street Transcript (2006). Available online: www.innocentive.com/about/media/200504_WSJ_Interview.pdf, (accessed 10 Jan 2006).

Webopedia-e-Government (2007) 'e-Government'. Available online: www.webopedia.com/TERM/E/e_government.html (accessed 5 July 2007).

Webopedia-Turing (2007) 'Turing test'. Available online: www.webopedia.com/TERM/T/Turing_test.html (accessed 5 July 2007).

Welfens, P. and Weske, M. (eds) (2006) *Digital Economic Dynamics: Innovations, Networks and Regulations*, Springer-Verlag.

Wenger, E. (1998) 'Communities of practice. Learning as a social system', *Systems Thinker*. Available online: www.co-i-l.com/coil/knowledge-garden/cop/lss.shtml, (accessed 5 September 2007).

Wenger, E. (1998) *Communities of Practice: Learning, Meaning, and Identity*, Cambridge MA: Cambridge University Press.

Wikipedia – Collective-Intelligence (2007) 'Collective intelligence'. Available online: http://en.wikipedia.org/wiki/Collective_intelligence (accessed 5 July 2007).

Wikipedia – Cyberspace (2007) 'Cyberspace'. Available online: http://en.wikipedia.org/wiki/Cyberspace (accessed 5 July 2007).

Wikipedia – Technology transfer (2007). Available online: http://en.wikipedia.org/wiki/Technology_transfer (accessed 10 September 2007).

Wikipedia – URENIO (2007). Available online: http://en.wikipedia.org/wiki/URENIO (accessed 10 October 2007).

Winters, N., Walker, K., and Roussos, G. (2007) 'Facilitating learning in an intelligent environment'. Available online: www2.theiet.org/oncomms/sector/computing/Library.cfm?ObjectID=821FC09A-0F2E-ABE9-31BC8B1023D13593 (accessed 21 August 2007).

Wolpert, D.H. and Tumer, K. (2001) 'An introduction to collective intelligence', NASA Tech Report. Available online: http://ic.arc.nasa.gov/ic/people/kagan/coin_pubs.html (accessed 5 May 2005).

Wood, P. (1991) 'Innovative cities in Europe', in J. Simmie (ed.), *Innovative Cities*, London: Spon Press, pp. 231–47.

Wooldridge, M. and Jennings, R.N. (1995) 'Intelligent agents: theory and practice', *The Knowledge Engineering Review*, 10 (2): pp. 15–152.

World Business (2007) 'Global innovation index: more on methodology'. Available online: www.worldbusinesslive.com/article/610009/global-innovation-index-methodology (accessed 28 August 2007).

Yorkshire Futures (2007). Available online: www.knowledge-rich.com (accessed 8 July 2007).

Zimmerman, R. and Horan, T. (eds) (2004) *Digital Infrastructures (Networked Cities)*, London: Routledge.

Index

Abramosky, L. *et al.* 25
absorption of technology 269–70
academia *see* universities
Almere 120
Anttiroiko, A.V. 61
'Archipelago Europe' study 30, 32–3
Argentina 52
artificial intelligence 17, 18, 19, 121–3
Asia 8, 12, 30, 41, 61, 137
Atlee, T. 122
augmented technology transfer 183, 192–4
  networking functionality 194
Austin 30
Australia 10, 52, 126, 129, 241
Austria 25, 42

Baden-Württemberg 14, 32, 33, 40, 80, 137
Bagnasco, A. 56
Bangalore 10, 30, 126, 129, 137
Barcelona 120, 243
Bario 125, 126, 130
Basque County 102, 153, 160
Baumgartner, Jeffrey 86
Bayern 14, 32, 33, 137
Becattini, G. 35, 58, 137
Beijing 30, 89
Belgium 25, 31, 160
Bell, Robert 243
benchmarking 131–4, 151–3
Booz Allen Hamilton 8–10
Borås 120
Borneo 130
Boston 30
Bowen-James, A. 112
Braunschweig 32
Brazil 8, 10, 52, 126

broadband networks 23–4, 88–9, 111–12, 116–19, 123–30, 213–14, 227–33, 245, 247, 253, 259, 267
brokering theories 78
Brussels 31
Buckinghamshire 31
Budapest 120
Bulgaria 25
business intelligence 141–3, 144

Calgary 125–8
California 30, 35, 116
Camagni, R. 59, 139
Cambridge 59, 178
Canada 52, 117, 125–8, 130
Canberra 241–2
Carroll, Darren J. 205
Catalonia 153
Cherukuri, Satyam 11
Chesbrough, H. 195
Cheshire 31
Chicago 230
China 8, 9, 10–11, 61, 111, 126–7, 129–30, 137, 230
Churchill, E. *et al.* 82
City-Vitals 132
Clausen, T.H. 51
Cleveland 126–8
closed innovation model 178, 195
clusters 15, 19, 35–41, 56–65, 79, 80, 140, 178
  actors in 58
  competitiveness 70
  content of 57
  innovation management within 60–1
  intelligence 143–6, 166–8
  and intelligent systems 72
  multi-cluster systems 67

Index

planned 61–2
and territorial systems 74–5
types of 57–8
virtual 85
within incubators 62–4
COINs 18–19
collaboration 41, 48, 61, 78, 90, 121, 143, 179, 195, 257–61, 265–6
collaborative innovation networks 18–19, 196–200
  actors in 197–9
  competences 199
  connectivity 199
  as intelligent space 222–5
  managing 205–22
  nodes within 201, 203–5, 224
  objectives 197
  partner autonomy/dependence 199
collaborative innovation platform 280, 281
collective creativity 35
collective intelligence 121–3, 143, 145, 160–1
collective learning 15, 59, 85, 139, 179
communication 8, 16, 180–1
Community Innovation Surveys (CIS) 21, 25, 29, 33, 41, 53
Community Support Framework 14
competitive advantage 8, 24, 35, 172
competitiveness clusters 70
Confidential Company Report 168–71
  data analysis 170–1
  data collection 169–70
  reporting 171
continuous spatial enlargement 56–73
Cooke, P. 37, 80, 138
cooperation 15, 35, 64
  networks 15, 181–3
Copenhagen 120
Cordis 10
  Technology Marketplace 154–5
Corporate Innovation Machine 86
'crowdsourcing' tools 86
cyber cities 113–15
  categories of 115
  principles of 114–15
  vs. intelligent communities 248
Cyberjaya 111
cyberspace 113–15
cycle of innovation 35, 82
Cyprus 25
Czech Republic 25, 26

Datta, A. and Mohtadi, H. 177
Davenport, T.H. and Prusak, L. 180
DECiLOR 146
Denmark 25, 52, 62, 211
developing countries 129–30, 137
developing regions 7, 67–9
digital cities 20, 88, 213–14, 226–46, 257–70, 275, 280
  and smart communities 112, 117
  sustainability of 245–6
digital divide 125
digital marketplaces 18
Digital Research Centre 183–92
  architecture 184
  data collection 185
  dissemination activities 189
  impact of 189
  obstacles 190
  online technology transfer platform 185
  R&D 190–1
  roadmaps 185–7, 191
digital spaces 42, 43, 44, 71–3, 81, 88
disruptive technologies 48
'district' theory 35–7
Dresner, Howard 141
Dundee 126

e-administration 243
e-city 117
  marketplaces 235–40
  promotion 240–2
e-commerce 18
e-communities 82
e-democracy 244
e-government 242–5
e-innovation tools 85–8
e-intelligence 82–3
e-learning 89–90, 183
e-marketplaces 88–9, 226–46
  platform 280, 283–4
  sustainability of 245–6
e-voting 244
East Midlands Observatory 147
East of England Observatory 147
economic intelligence 140–1
Edison, Thomas A. 78
Edquist, C. et al. 139
Edvinsson, L. 17
Eindhoven 39
Ennis 126, 130
Enright, M. 40–1

entrepreneurialism 16, 24, 49
environment
　for innovation 16, 35, 41, 52
　relationship between external and internal 51
EPINETTE 146
Estonia 25, 89, 95, 102, 120, 126
Etelä-Suomi 32
Europe 7, 8, 10, 11–13, 20–44, 61, 65, 73, 111, 195
　core of excellence 33–8
　　making the core 38–44
　innovation performance in states 27
　innovation policy 65–7
　and intelligent cities 119–20
　regional disparities in innovation performance 28
　regional inequalities of Knowledge 14
　regions of innovation 21, 31
　rise of the innovation economy in 23–9
European Competitiveness Reports 29
European Innovation Scoreboards 22, 24–5, 29, 31, 32, 66, 75, 132, 153, 211
European Regional Development Fund 21, 148, 153
European Research Area 12
European Trend Chart on Innovation 22, 31
exchange of information 8, 105, 140
explicit learning 89–90
Extremadura 95, 103, 160

FAST study 30
Finland 13, 22, 25, 26, 31, 32, 33, 43, 44, 120, 137, 153, 211
Flat World 10
Florida 119, 122, 126–8, 132
Fogel, D.B. 121
Ford, Henry 78–9
France 14, 25, 31–3, 52, 70, 120, 126, 137, 145–6, 153
Freeman, C. 52, 197
Friedman, Thomas 10

Gangnam 125–7
Geneva 244
Germany 14, 25, 31–3, 40, 42, 44, 52, 80, 95, 106, 120, 137
Gibson, William 113
Glasgow 119, 125–7

global
　innovation index 28
　innovation networks 1, 3, 16, 43, 247
　metropoles 125–7
　promotion 270
globalisation 7–20, 137
'globalisation of innovation' 11, 20, 137, 196
Gloor, P. 19
Goldense, B. and Schwartz, A. 195
governance 41, 56, 73–6
Greece 25, 71, 148, 153, 160, 192, 208, 211, 248
Griliches, Z. 50, 54

Hagedoorn, J. et al. 197
Hamburg 120
Hanover-Brunswick-Gottingen 80
Hansson, F. et al. 62
Hargadon, A. 78
Helsinki 30, 43, 120, 174, 229, 259
Henri, F. and Pudelko, B. 84
Holmes, J.S. and Glass, J.T. 216–17
Hong Kong 89
Howells, J. 204
human-centred economy 8,11
human resources 8, 29, 31, 42
Hungary 25, 120

Ichikawa 125–7
Ile-de-France 14, 31, 32, 33, 178
incremental innovations 48
incubators 62–4
in-house innovation model 34
India 8, 9, 10–11, 126, 137, 230
'industrial district' theory 35–7
information exchange 8, 105, 140
InnoCentive 205–7, 265
innovation 8–20, 21–45, 47–9
　capability 247
　and cities 15
　cycle of 35, 82
　definition of 47
　as an environmental condition 77–80
　EU regional disparities in performance 28
　externalisation of 139
　gap between US and EU 11–12
　general purpose 48
　globalisation of 11, 20, 137, 196
　growth engine 138
　incremental 48

and intelligence 121
and intelligent environments 119–20
Islands of 30, 67
management within clusters 60
management technologies (IMTs) 91–3
obstacles to 53
organisational 47–8
performance in EU states 27
process 47
product 47
radical 48
scoreboards *see* European Innovation Scoreboards
stimulating 12
systems of 45–76, 79
threshold to 175
through collaboration 195–6
tools 73
and virtual space 106–9
innovation systems 54–6
analysis 59–60
criticism of 79
elements 54
governance 56
relationships 54–6
innovative regions 137–40
intelligence 120–3, 140–1
artificial 17, 18, 19, 121–3, 157
business 141–3, 144, 168–71
collective 121–3, 143, 145, 160–1
cluster 143–6, 166–8
distributed 156–9
and integration 157–9
regional 144, 147–9
strategic economic 137–73, 263–4, 268–9
intelligent cities 19–20, 43, 110–34, 247–84
architecture of 267–74
around the world 125–31
characteristics 111–12
creating 275–84
cyber cities 113–15
defining 120–3
as digital spaces of collaboration 257–61
environments 17–20
knowledge functions of 261–7
metrics 131–4
platforms for 275–84
shaping 247–9
smart communities 116–17
as territorial systems of innovation 249–57
three layers of 123–4
a window to global networks 270–4
intelligent
communities 117–19, 248
Community Forum 111, 112, 117–18, 125, 249, 267
environments 16–20
global systems of innovation 71–3
island 127
Room 112
space 222–5
virtual environments 18
Ireland 25, 42, 126, 130, 211
Ishida, T. 159
Islands of Innovation 30, 67
Israel 30, 52
Issy-les-Moulineux 125, 126
Italy 23, 25, 33, 35, 52, 56–7, 59, 60, 86, 120, 211

Jaffe, A.B. 49–50, 59
Japan 10, 11, 26, 30, 52, 61, 79, 86, 125–8, 137, 195, 200
Johnson, A. 143

Karlsruhe 31, 32
Kaufmann, A. and Todtling, F. 249
Keeble, D. *et al.* 139
Kennedy, J. and Eberhart, R. 157
Kessler, E. 205
Kista 137
knowledge 8, 14, 15, 16, 78
-based economy 7–8, 11, 13, 21–5, 37, 45, 128, 153
evolution 55
flows 250
-intensive clusters 44, 267–8
management 43, 80
tools 79, 140
networks 139, 250–7, 268
Rich Programme (KRICH) 148
spillovers 15, 59, 103
tacit to explicit 64, 79, 89–90, 250
Korea 52, 111
Krugman, P. 37
Kyoto 259

LaGrange 125, 126, 130
Landabaso, M. 17, 138
Lawson, C. and Lorenz, E. 59, 64

learning process 81
'learning region' theory 35, 37, 64, 65
  criticism of 38
Lévy, Pierre 114–15, 145
licensing 175–6
Limousin 153
Lisbon Strategy 12–13, 24, 29, 153, 208
Little, Arthur D. 261–3
'Living Labs' 120
London 13, 33, 120
Lorenz, E. 139
Lorraine 95, 145–6
Lundvall, B. 37, 138, 249
Luxembourg 26
Lyon 153

Macedonia 103, 153, 160, 211
Madrid 32
Madri+d 155–6
Malaysia 111, 125, 126, 130, 137
Malta 25
managing networks 195–225
  collaborative innovation as intelligent
    space 222–5
  contact vs. cost 202–4
  digital spatiality and networking 204–5
  InnoCentive 205–7
  new product development based on
    vertical networks 216–22
  NPD-Net 217–22
  Regional Innovation Poles 208–16
  selection of partners 200–2
  spatiality of collaboration 202–4
  trust vs. competence 200–2
Manchester 127–8
Manhattan 125
mapping 72, 132
Markusen, A. 41
Marshall, G.L. 37, 107
Maskell, P. and Malmberg, A. 37
Massachusetts 30, 35
  Innovation Economy Index 152
Mataró 120
Mazower, M. 248
Meta-foresight 160–72
  data analysis 165
  data collection 163–4
  dissemination 165–6
  platform 161–6
  sectoral/cluster intelligence 166–8
  structure 162
Millson, M.R. and Raj, S.P. 201

Milton Keynes 178
Minsky, M. 157
Mitaka 125, 126
modelling 131–4
Morgan, K. 79, 107
Multi-cluster regional systems 41–2
multi-cluster systems 67
Munich 30

Narula, R. and Hagedoorn, J. 199–200
Nelson, R. 249
Nelson, R. and Rosenberg, N. 52, 56
Nelson, R. and Winters, S. 37, 51, 80
Netherlands 13, 25, 31, 32, 33, 39, 120,
  137, 211, 228
Nevada 126, 130
new company incubation platform 280,
  282
new product development 216–22, 251–3,
  265–6
New Songdo City 89
New York 30, 89, 119, 125, 126
Nonaka, I. and Takeuchi, H. 79, 139
Noord-Brabant 13, 31, 32, 33, 39, 137
Nordrhein-Westphalen 103
North, K. and Kares, S. 132
North Carolina 30
Norway 51
Novak, M. 112
NPD-Net 217–22

Oberbayern 31, 32
Ohio 127
Ontario 125
open innovation paradigm 80, 178, 195
organisational innovation 47–8, 254–7
Oslo Innovation Manual 29, 47, 75
Östra Mellansverige 32
outsourcing 9, 11, 88, 140, 195
Oxfordshire 31
Ozer, M. 204

Paci, R. and Usai, S. 59
patents 26, 32, 50
Piergiovanni, R. and Santarelli, E. 50
Piore, M. and Sabel, C. 35
Pirai 125, 126
Pittaway, L. *et al.* 197
planned clusters 61–2
platforms for intelligent cities 275–84
  collaborative innovation platform
    280–1

new company incubation platform 280, 282
strategic economic intelligence platform 275–7
technology dissemination platform 277–9
virtual tour and e-marketplace platform 280, 283–4
Pohjois-Suomi 32
Poland 25
population-intense countries 8, 11
Porter, M. 37, 56–7, 137
Portugal 25, 211
Poti, B. and Basile, R. 59
precarious regional systems of innovation 67–9
process innovations 47, 253–4
Programmes of Regional Innovative Actions 65
Putnam, R. 122
Putrajaya 111

R&D 7–20, 26, 32–3, 42–4, 49, 59, 81, 195
  expenditure on 10, 32
  in-house 49
  international R&D alliances 196
  spillover 50
  of universities 49–50
Radjou, N. 198
Radovanovic, D. 248
Ragusa 17, 248
Raleigh 30
regional
  Documentation Centres 149
  Innovation and Technology Transfer Infrastructures and Strategies 14, 65
  Innovation Poles 69–71, 208–16
  Innovation Strategies 14, 65–6
  intelligence 144, 147–9
  Intelligence Unit 148
  Observatories 147–8
  policy periodic reports 21
  systems of innovation 33–8, 64–71
  Technology Plans 65
regions of innovation excellence 21–44
research and development see R&D
Rhone-Alpes 14, 31, 33, 137
Romania 25
Russia 8, 10–11, 86

San Francisco 89
Sant Cugat 120
Santos, J. and Doz, Y. and Williamson, P. 196
Saxenian, A. 35
Saxony 80
Schumpter, Josef 49, 60, 249
Schwen, T. and Hara, N. 81
science parks 61–2
Scotland 119, 125–7, 244
Scott 35, 37, 137
Seattle 30
selection of partners 200–2
Seoul 30, 89, 111, 119, 125–7
Shanghai 30
Silicon Alley, Manhattan 125
Silicon Valley 30, 137, 138, 174
Singapore 10, 89, 111, 119, 125–7, 137
Slovakia 25
smart cities *see* intelligent cities
smart communities 116–17
Smeds, R. and Alvesalo, J. 82
social capital 17, 18, 27, 62, 122
Songdo 111
Sophia-Antipolis 120, 137
South Korea 10, 126
Spain 25, 32, 95, 120, 153, 160, 211
spatial competitive advantage 35–7, 39
Spokane 111, 126
Sternberg, R. 80
Stockholm 13, 30, 31, 32, 33, 89, 137
Storper, M. 138
Storper, M. and Scott, A. 35, 37
strategic business intelligence 168–72
strategic economic intelligence 137–73, 263–4, 268–9
  main components of 149–56
    competition benchmarking 151–3
    foresight 153–4
    market and technology watch 150–1
    Meta-foresight 160–72
    the 'next big thing' 172–3
    platform 275–7
    R&D watch 154–6
strategic information 140
Stuttgart 31, 32, 40, 120, 174
Summary Innovation Index 11, 25, 66
Sunderland 126–8
supply chains 7, 15, 43, 71, 196, 216, 254, 256
Sweden 13, 22, 25, 26, 31, 32, 33, 42, 52, 120, 137

Sydsverige 32
systemic theories of innovation 16, 50, 64, 79, 249
systems of innovation 45–76, 79
Szuba, T. 122

Taipei 111, 119, 125–7, 129, 137
Taiwan 10, 30, 52, 125–7
Tallinn 89, 120, 126
technological regimes 51–2
technology dissemination platform 277–9
technology parks 61–2
technology transfer 174–94, 264
　augmented 183, 192–4
　as communication 180–1
　Digital Research Centre 183–92
　and knowledge networks 255
　North-South 177
　routes 175–7
　and universities 177–80
Teheran Valley 127
telecommunications 23, 42–4, 57, 80, 111, 128, 155, 204, 210, 213, 227, 232, 274
territorial systems of innovation 2, 3, 20, 38, 46, 72–6, 113, 119, 123–4, 137–8, 143, 179, 261, 266–7
　advantages and challenges of 75
　and intelligent cities 249–57
Thailand 10
Thessaloniki 92, 160, 183, 192, 209, 211, 248
Thessaly 103, 148
Tianjin 125, 126, 129–30
Tokyo 30, 89, 126
Torino 120
Toronto 126
Toulouse 174
Turing Test 121
Tuscany 59
Tyrol 102, 106

United Kingdom 13, 25, 31, 33, 42, 52, 62, 120, 125–8, 147–8, 153, 195
United States 8, 10–12, 26, 30, 32, 35, 41, 52, 61, 86, 111, 117, 119, 125–7, 130, 137, 152, 211, 236

universities 15, 53, 57, 59, 128, 183, 196
　and technology transfer 175–80
Utrecht 31
Uusimma 13, 31, 32, 33, 43, 137, 153

Västervik 120
Västsverige 32
VERITE 77, 91–109
　building blocks of 92
　conclusions 106–9
　direct impact 102–6
　discussion forum 99
　goals of 91
　knowledge base 98–9
　language 105–6
　learning and the virtual space 104–6
　learning impact 100–1
　motivation for learning 101
　services and tools 97–8, 108
　and understanding 108
　as virtual community 95–100, 106–9
Victoria 126, 129
virtual
　clusters 85
　communities 84–5
　　VERITE 95–100
　innovation environments (VIEs) 77–109, 268
　networks 71–3
　and physical worlds 18
　spaces 181–3

Wales 95, 160, 223
Waterloo 125, 126
Weber, Alfred 37
Wenger, E. 84
Western Valley 126, 130
Wood, P. 37
World Foundation for Smart Communities 112, 116
World Teleport Association 117

Yet2.com 155
Yokosuka 119, 126–8
Yorkshire Futures 147–8

Zuid-Holland 13, 31, 33

eBooks – at www.eBookstore.tandf.co.uk

A library at your fingertips!

eBooks are electronic versions of printed books. You can store them on your PC/laptop or browse them online.

They have advantages for anyone needing rapid access to a wide variety of published, copyright information.

eBooks can help your research by enabling you to bookmark chapters, annotate text and use instant searches to find specific words or phrases. Several eBook files would fit on even a small laptop or PDA.

NEW: Save money by eSubscribing: cheap, online access to any eBook for as long as you need it.

Annual subscription packages

We now offer special low-cost bulk subscriptions to packages of eBooks in certain subject areas. These are available to libraries or to individuals.

For more information please contact webmaster.ebooks@tandf.co.uk

We're continually developing the eBook concept, so keep up to date by visiting the website.

www.eBookstore.tandf.co.uk